SpringerBriefs in Mathematics

W0230278

SpringerBriefs in Mathematics showcases expositions in all areas of mathematics and applied mathematics. Manuscripts presenting new results or a single new result in a classical field, new field, or an emerging topic, applications, or bridges between new results and already published works, are encouraged. The series is intended for mathematicians and applied mathematicians.

More information about this series at http://www.springer.com/series/10030

SBMAC SpringerBriefs

The **SBMAC SpringerBriefs** series publishes relevant contributions in the fields of applied and computational mathematics, mathematics, scientific computing, and related areas. Featuring compact volumes of 50 to 125 pages, the series covers a range of content from professional to academic.

The Sociedade Brasileira de Matemática Aplicada e Computacional (Brazilian Society of Computational and Applied Mathematics, SBMAC) is a professional association focused on computational and industrial applied mathematics. The society is active in furthering the development of mathematics and its applications in scientific, technological, and industrial fields. The SBMAC has helped to develop the applications of mathematics in science, technology, and industry, to encourage the development and implementation of effective methods and mathematical techniques for the benefit of science and technology, and to promote the exchange of ideas and information between the diverse areas of application.

http://www.sbmac.org.br/

Alexandre L. Madureira

Numerical Methods and Analysis of Multiscale Problems

 Springer

Alexandre L. Madureira
Laboratório Nacional de Computação Científica-LNCC
Petrópolis, Brazil

Fundação Getúlio Vargas-FGV
Rio de Janeiro, Brazil

ISSN 2191-8198   ISSN 2191-8201   (electronic)
SpringerBriefs in Mathematics
ISBN 978-3-319-50864-1   ISBN 978-3-319-50866-5   (eBook)
DOI 10.1007/978-3-319-50866-5

Library of Congress Control Number: 2017931654

Printed on acid-free paper

This Springer imprint is published by Springer Nature
The registered company is Springer International Publishing AG
The registered company address is: Gewerbestrasse 11, 6330 Cham, Switzerland

Preface

Multiscale problems are omnipresent in real-world applications and pose a challenge in terms of numerical approximations. Well-known examples include modeling of plates and shells, composites, neuronal modeling, and flows in porous media. Physically, what characterize such problems is the presence of different important physical scales. Mathematically, it means that derivatives of the solution might blow up as some parameter goes to zero, as in singular perturbed problems, for instance.

Accordingly, the partial differential equations (PDEs) that model these problems are characterized by either the presence of a small parameter in the equation (e.g., the viscosity of a turbulent flow) or in the domain itself (as in shell or neuroscience problems). Denote this small parameter by ε. Frequently, there is interest in investigating or approximating the solutions for positive but nonvanishing ε. However, they might behave as in the $\varepsilon = 0$ case or, even more interestingly, as in an intermediate, asymptotic state, which depends on the problem under consideration. It hardly comes as a surprise that, in general, standard numerical methods do not perform well under all regimes.

It is then important to discuss modeling of multiscale PDEs, where modeling has two meanings. It can be in the sense of approximating the original PDE by other equations that are easier to solve, as in plates or domains with rough boundaries. It can also be in numerical approximation point of view, where the final goal is to develop a numerical scheme that is robust, i.e., that works well for a wide range of parameters.

Understanding such asymptotic behaviors for different problems, and designing and analyzing robust models and numerical schemes, is the goal of these notes. As much as possible, the problems are considered in simple settings, so that technical details do not hinder the understanding of what is essential in each case considered.

The techniques involved are introduced by means of case studies, and I derive modeling error estimates by means of asymptotic analysis. The problems I describe involve equations with reaction, advection, and diffusion terms and problems with oscillatory coefficients or posed in domain with rough boundaries (like in a golf ball). I also introduce, in a general setting, some finite element methods that are suitable to deal with multiscale problems.

Prerequisites and Contents

This text is mainly oriented toward advanced undergraduate and graduate students. It might be also useful to researchers that are willing to read a bit about multiscale problem without digging too deep into technical details or sophisticated problems, as in research papers. I assume basic knowledge of analysis, mainly regarding Sobolev spaces, and some topics on functional analysis. It is however not essential that the reader completely master these results to follow the present text, and proper references are mentioned when needed. It would also be useful to have some experience with finite element methods, but the main tools are developed in these notes.

Chapter 1 introduces some basic notation and results that are useful throughout the book. It also contains a general description of several important finite element methods. Next, in Chap. 2, I consider one-dimensional singular perturbed reaction–advection–diffusion problems, in terms of numerical discretization and asymptotics. In Chap. 3, I discuss a one-dimensional equation of interest in computational neuroscience. It is a diffusion–reaction problem, with Dirac deltas (modeling the presence of synapses in a neuron) introducing layers in the interior of the domain. I show how a multiscale method can tackle the problem. A two-dimensional reaction–diffusion problem is considered in Chap. 4, and there I discuss how to develop the boundary layer terms for two-dimensional domains and introduce a multiscale finite element method based on enriching finite element spaces. Chapter 5 concerns PDEs posed in domains with rough boundaries, and I develop asymptotic expansions of the solutions and propose and analyze a finite element method of multiscale type. Finally, Chap. 6 deals with the classical problem of elliptic PDEs with rough coefficient. I develop its asymptotic expansion and analyze a numerical method in a simple one-dimensional setting.

Acknowledgments

Of course this book would never be possible without many enlightening iterations with wonderful researchers. I was introduced to the field of numerical analysis of multiscale problems by the late Leopoldo Franca and then proceeded to work under the guidance of Doug Arnold. Both of them fundamentally shaped my career. The influence of my longtime friend and main collaborator Frédéric Valentin is of order ε^{-1}, and the choice of topics in the book reflects that. He also read the manuscript and offered suggestions that improved the notes.

Parts of this book were drawn from [184, Chap. 4] and short courses delivered at Denver University, USA; Universitá degli Studi di Pavia, Italy; and Brazilian Universities, and I thank Leo Franca, Daniele Boffi, Carlo Lovadina, Paulo Bösing, Igor Mozolevski, and Sandra Malta for the kind invitations. Several other people influenced in manifold ways the outcome of the book. I am particularly grateful to my colleagues at LNCC, students, and friends. The publishing of this book under the SpringerBriefs seal would never come to life without the relentless support and encouragement of my editor and friend Mariano Carvalho.

This book was completed while I was spending a sabbatical year at the Division of Applied Mathematics at Brown University. During such period, I was lucky enough to learn from my host Johnny Gúzman and also Marcus Sarkis from WPI. Their wit and hospitality and the general atmosphere of the Division confirmed that top-notch mathematics and friendliness are definitely compatible.

I gratefully acknowledge the hospitality of the Division of Applied Mathematics at Brown University and the long-term financial support of the Brazilian funding agencies CNPq and FAPERJ.

I dedicate this work to my wife Daniele and daughter Maria. They make life wonderful.

Petrópolis and Rio de Janeiro, Brazil   Alexandre L. Madureira

Contents

Chapter 1
Introductory Material and Finite Element Methods

Abstract In this chapter, we introduce some notation, and also state some basic results regarding the Galerkin Method. In particular, some elementary estimates are presented, highlighting the importance of coercivity constants. This chapter also contains a brief introduction to some alternative methods, such as the *Residual Free Bubble Method*, the *Multiscale Finite Element Method*, the *Localized Orthogonal Decomposition*, the *Variational Multiscale Method*, and *Hybrid* and *Stabilized Methods*.

1.1 Introduction

We consider here introductory concepts of weak formulations of PDEs and Sobolev spaces. Several books deal with these subjects from different points of view [20, 22, 38, 54, 65, 146, 155].

For the sake of simplicity, whenever possible we deal with one-dimensional domains. But that is not always feasible. For instance, when dealing with domains with rough boundaries, the simplest case is two-dimensional. We also consider two-dimensional domains when we want to introduce some new techniques, as in the case of boundary-fitted coordinates. In these cases, $\Omega \in \mathbb{R}^2$ is an open bounded set. A typical point $\mathbf{x} \in \Omega$ is written as $\mathbf{x} = (x_1, x_2)$, and, in general, two-dimensional vector is in bold. We often impose that the boundary $\partial\Omega$ is smooth, and that means that the boundary is locally a graph of a C^∞ function. In other instances, more general domains are important, and a Lipschitz domain is considered, with the meaning that its boundary is locally a graph of Lipschitz function—the typical example to have in mind are polygons. In any case, cuspids are not allowed. See [54, 116, 170] for precise definitions.

We introduce next some Sobolev spaces related to two-dimensional domains. Similar definitions hold for other dimensions, mutatis mutandis.

For $1 \le p < \infty$, let $L^p(\Omega)$ be the space of measurable functions $v : \Omega \to \mathbb{R}$ such that $|v|^p$ is Lebesgue integrable, endowed with the norm

$$\|v\|_{L^p(\Omega)} = \left(\int_\Omega |v(\mathbf{x})|^p \, d\mathbf{x} \right)^{1/p}.$$

© The Author(s) 2017
A.L. Madureira, *Numerical Methods and Analysis of Multiscale Problems*,
SpringerBriefs in Mathematics, DOI 10.1007/978-3-319-50866-5_1

We also define the space of *essentially bounded functions*:

$$L^\infty(\Omega) = \{v : \Omega \to \mathbb{R} : \|v\|_{L^\infty(\Omega)} < \infty\}, \qquad \|v\|_{L^\infty(\Omega)} = \operatorname*{ess\,sup}_{\mathbf{x} \in \Omega} |v(\mathbf{x})|.$$

We define $H^k(\Omega)$ as the set of functions $v : \Omega \to \mathbb{R}$ such that all weak derivatives

$$\frac{\partial^{k_1+k_2} v}{\partial x_1^{k_1} \partial x_2^{k_2}} \in L^2(\Omega),$$

for all non-negative integer k_1, k_2 such that $k_1 + k_2 \le k$. Using the semi-norms $|v|_{H^k(\Omega)}$, where

$$|v|_{H^k(\Omega)}^2 = \sum_{\substack{k_1,k_2=0 \\ k_1+k_2=k}}^{k} \left\| \frac{\partial^{k_1+k_2} v}{\partial x_1^{k_1} \partial x_2^{k_2}} \right\|_{L^2(\Omega)}^2,$$

we endow $H^k(\Omega)$ with the norm

$$\|v\|_{H^k(\Omega)} = \left(\sum_{i=0}^{k} |v|_{H^i(\Omega)}^2 \right)^{1/2}.$$

For $k = 0$ we write $H^0(\Omega) = L^2(\Omega)$. A property that we use often is that one-dimensional functions in H^1 or two-dimensional functions in H^2 are continuous. Finally, we define $H_0^k(\Omega)$ as the closure of $C_0^\infty(\Omega)$ (space of infinitely differentiable functions with compact support in Ω) in the norm $\| \cdot \|_{H^k(\Omega)}$.

The result below establishes an important inequality [34, 92, 116].

Lemma 1.1 (Poincaré's Inequality) *Let D be a one- or two-dimensional bounded domain, with Lipschitz boundary. Then there exists a constant c that depends on D such that*

$$\|u\|_{L^2(D)} \le c \| \nabla u\|_{L^2(D)}.$$

for all $u \in H_0^1(D)$.

Throughout these notes, we adopt the convenient convention that the constants that appear in estimates are not the same everywhere, and are independent of the parameters ε and h, unless explicitly indicated. They often depend on the domains and f, and are generally denoted by c.

The problems considered in this text are of variational nature and are set in Hilbert spaces [38, 55, 152, 154]. So, let V be such a space, with norm $\| \cdot \|_V$. Consider the bilinear form $a(\cdot, \cdot) : V \times V \to \mathbb{R}$, and $f \in V^*$, the dual space of V. Consider now $u \in V$ such that

$$a(u, v) = \langle f, v \rangle \quad \text{for all } v \in V. \tag{1.1}$$

The symbol $\langle f, v \rangle$ denotes the action of f on v, and $\|f\|_{V^*} = \sup_{v \in V} \langle f, v \rangle / \|v\|_V$ is the *operator norm* (the convention here is that the supremum is considered in $V \setminus \{0\}$). Thus $\langle f, v \rangle \leq \|f\|_{V^*} \|v\|_V$ for all $v \in V$.

In our case, (1.1) is related to a differential operator $\mathscr{L} : V \to V^*$ such that $\langle \mathscr{L} u, v \rangle = a(u, v)$ for all $v \in V$. So, in general, the statement

$$\mathscr{L} u = f. \tag{1.2}$$

should be actually understood as (1.1). However under some smoothness assumptions, (1.2) can be also interpreted in the *pointwise sense*, and we say that the *weak form* (1.1) is equivalent to the *strong form* (1.2).

Example 1.1 Consider the problem of finding a function $u : [0, 1] \to \mathbb{R}$ such that

$$- (\varepsilon u')' + bu' + cu = f \quad \text{in } (0, 1), \qquad u(0) = u(1) = 0, \tag{1.3}$$

where ε, b, c, f are all real, "smooth enough" real functions in $(0, 1)$. Multiplying both sides of the above equation by an arbitrary $v \in H_0^1(0, 1)$ and integrating in $(0, 1)$, we gather that $u \in H_0^1(0, 1)$ solves

$$\int_0^1 \varepsilon u' v' \, dx + \int_0^1 bu' v \, dx + \int_0^1 cuv \, dx = \int_0^1 fv \, dx \quad \text{for all } v \in H_0^1(0, 1). \tag{1.4}$$

Note the integration by parts in the first term reduces the regularity requirements of the functions involved. If we define $a(u, v) = \int_0^1 \varepsilon u' v' \, dx + \int_0^1 bu' v \, dx + \int_0^1 cuv \, dx$, and $\langle f, v \rangle = \int_0^1 fv \, dx$, then (1.4) is (1.1). Also, if $\mathscr{L} u = (\varepsilon u')' + bu' + cu$, then (1.3) is (1.2).

Remark 1.1 A stylistic remark is that we use both d/dx and $'$ to denote one-dimensional derivatives interchangeably.

Example 1.2 For a given $f \in L^2(0, 1)$, let $u^\varepsilon \in H_0^1(0, 1)$ be the weak solution of

$$-\frac{d}{dx} \left[a^\varepsilon(x) \frac{d}{dx} u^\varepsilon(x) \right] = f(x) \quad \text{in } (0, 1),$$
$$u^\varepsilon(0) = u^\varepsilon(1) = 0. \tag{1.5}$$

For each $\varepsilon > 0$, consider $a^\varepsilon \in L^\infty(0, 1)$. Moreover, assume that there are constants β and α that are ε-independent and such that $\beta \geq a^\varepsilon(x) \geq \alpha > 0$ for almost all $x \in (0, 1)$. Multiplying both sides of the equation by an arbitrary $v \in H_0^1(0, 1)$ and integrating in $(0, 1)$ by parts, it results that

$$\int_0^1 a^\varepsilon(x) \frac{du^\varepsilon}{dx}(x) \frac{dv}{dx}(x) \, dx = \int_0^1 f(x) v(x) \, dx \qquad \text{for all } v \in H_0^1(0, 1). \tag{1.6}$$

Two properties are sufficient to guarantee existence and uniqueness of solutions for (1.1). These are *coercivity* (a.k.a. ellipticity or stability) and *continuity* of the bilinear form. Coercivity holds if there is a positive constant α such that

$$a(v, v) \geq \alpha \|v\|_V^2 \quad \text{for all } v \in V. \tag{1.7}$$

The bilinear form is continuous if there exists a constant c such that

$$a(u, v) \leq c\|u\|_V \|v\|_V \quad \text{for all } u, v \in V. \tag{1.8}$$

The Lax–Milgram Lemma below establishes existence and uniqueness for (1.1). It is not in the most general form [23], but is enough for many problems of interest [22, 31, 34, 92].

Lemma 1.2 (Lax–Milgram) *Let V be a Hilbert space, $a(\cdot, \cdot) : V \times V \to \mathbb{R}$ a coercive (1.7) and continuous (1.8) bilinear form, and $f \in V^*$. Then (1.1) has a unique solution. Moreover, $\|u\|_V \leq (1/\alpha)\|f\|_{V^*}$.*

Proof If $a(\cdot, \cdot)$ is symmetric, then the Lax–Milgram Lemma is a simple application of the Riesz representation theorem, as $a(\cdot, \cdot)$ defines an inner product in V. The proof for the general case is presented in any good finite element book, for instance in the references cited above.

The estimate follows from

$$\alpha \|u\|_V^2 \leq a(u, u) = \langle f, u \rangle \leq \|f\|_{V^*} \|u\|_V. \qquad \square$$

1.2 Galerkin Method

The Galerkin approximation method for (1.1) consists in choosing a closed subspace $V_h \subset V$, usually of finite dimension, and search for $u_h \in V_h$ such that

$$a(u_h, v_h) = \langle f, v_h \rangle \quad \text{for all } v_h \in V_h. \tag{1.9}$$

Since V_h itself is a Hilbert space, existence and uniqueness of solutions for (1.9) follow immediately from Lax–Milgram Lemma 1.2, if $a(\cdot, \cdot)$ is coercive and continuous.

A first error estimate is simple to obtain. First note the *Galerkin orthogonality*

$$a(u - u_h, v_h) = a(u, v_h) - a(u_h, v_h) = \langle f, v_h \rangle - \langle f, v_h \rangle = 0 \quad \text{for all } v_h \in V_h.$$

The result below is fundamental in establishing error estimates.

Lemma 1.3 (Céa) *Assume the bilinear form $a(\cdot, \cdot)$ to be coercive with coercivity constant α, and continuous with continuity constant β. Let $u \in V$ be the solution* *of* (1.1), *and* $u_h \in V_h$ *solution of* (1.9). *Then*

$$\|u - u_h\|_V \le \frac{\beta}{\alpha}\|u - v_h\|_V \quad \text{for all} v_h \in V_h.$$

Proof Let $v_h \in V_h$. Then it follows from the coercivity, Galerkin orthogonality, and continuity that

$$\alpha\|u - u_h\|_V^2 \le a(u - u_h, u - u_h) = a(u - u_h, u) = a(u - u_h, u - v_h)$$

$$\le \beta\|u - u_h\|_V\|u - v^h\|_V \quad \text{for all } v_h \in V_h. \quad \square$$

The above lemma conveys an important information, that is, the Galerkin solution is the best approximation in V_h to the exact solution, up to a constant. That does not mean that u_h is a good approximation for u. That all depends on how well functions in V can be approximated by functions in the space V_h, and also on how large is β/α.

Whenever $a(\cdot, \cdot)$ is symmetric, it defines an inner product and the related *energy norm* $\|\cdot\|_E = a(\cdot, \cdot)^{1/2}$ in V. It follows from the Cauchy–Schwartz inequality that the estimate of Lemma 1.3 reduces to

$$\|u - u_h\|_E \le \inf_{v_h \in V_h} \|u - v^h\|_E. \tag{1.10}$$

In this instance, the Galerkin method can be seen as an energy minimizer, and the method is often called *Ritz Method* [22, 34, 148, 178].

Considering the use of finite elements, a fundamental step of the method is the discretization of the domain. Consider the one-dimensional domain $(0, L)$ and $V = H_0^1(0, L)$, for instance. Then it is very simple to introduce a discretization into finite elements by defining the nodal points

$$0 = x_0 < x_1 < \cdots < x_{N+1} = L, \qquad x_j = jL/(N + 1). \tag{1.11}$$

and the mesh parameter $h = L/(N + 1)$. Next, we define the finite dimensional space $V_h \subset V$ as $V_h = P_1$, where

$$P_1 = \{v_h \in V : v_h \text{ is linear in } (x_{j-1}, x_j) \text{ for } j = 1, \ldots, N + 1\}. \tag{1.12}$$

We say that V_h is a space of continuous piecewise linear functions, and a typical function of V_h is depicted in Fig. 1.1.

Observe that a function in V_h is completely characterized by the values it assumes at the nodal points x_1, x_2, etc. We introduce then a basis for V_h defined by $\phi_i \in V_h$ such that $\phi_i(x_j) = \delta_{ij}$ for $j = 1, \ldots, N$, where

$$\delta_{ij} = \begin{cases} 1 & \text{if } i = j, \\ 0 & \text{if } i \ne j, \end{cases}$$

and a typical basis function is represented in Fig. 1.2. Then $V_h = \text{span}\{\phi_1, \ldots, \phi_N\}$.

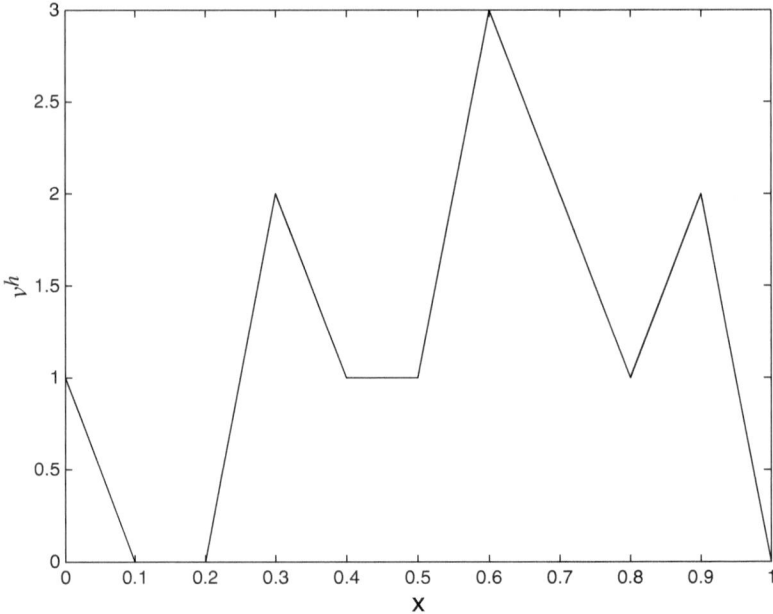

Fig. 1.1 Typical continuous piecewise linear function in P_1, as defined by (1.12). Notice that the function is completely characterized by its nodal values

Fig. 1.2 A basis function of the finite element space P_1 defined by (1.12)

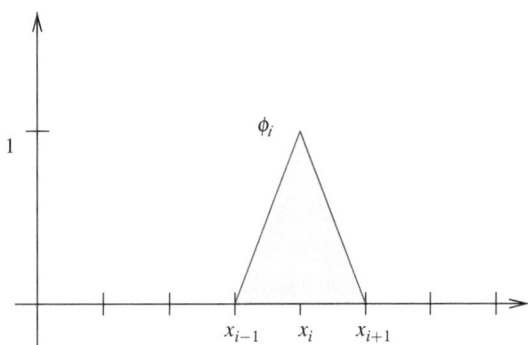

Finally, if $u_h(x) = \sum_{i=1}^{N} u_i \phi_i(x)$, we rewrite (1.9) as

$$\sum_{i=1}^{N} u_i a(\phi_i, \phi_j) = \langle f, \phi_j \rangle \quad \text{for } j = 1, \dots, N. \tag{1.13}$$

Note that $u_j = u_h(x_j)$ is the unknown value of u_h at the node x_j. In matrix form this consists in finding the $N \times 1$ vector $\mathbf{u} = [u_1, \dots, u_N]^T$ such that

$$\mathbf{M}\mathbf{u} = \mathbf{f},$$

where the $N \times N$ matrix $\mathbf{M} = (M_{i,j})$ and the $N \times 1$ vector $\mathbf{f} = [f_1, \ldots, f_N]^T$ are given by

$$M_{i,j} = a(\phi_i, \phi_j), \qquad f_j = \langle f, \phi_j \rangle.$$

For two-dimensional domains, the procedure is basically the same, but we discretize Ω in triangles or quadrilaterals, which we call *elements* and denote by K. For simplicity, we consider triangular elements only, unless noted otherwise, and denote the set of triangles by $\mathcal{T}_h$. Such partition of Ω cannot be arbitrary, and hanging nodes or overlaps are not allowed. Then V_h would be the space of piecewise linear continuous functions associated with this triangulation, i.e. continuous functions that are linear inside each triangle. This naturally generalizes (1.12) for higher dimensions, i.e., when dealing with two-dimensional spaces, we let

$$P_1 = \{v_h \in V : v_h|_K \text{ is linear for all triangles } K \in \mathcal{T}_h\}. \tag{1.14}$$

In both one and two dimensions, standard interpolation estimates yields [22, 54, 140, 146]

$$\|u - I^h u\|_{H^1(0,1)} \le ch|u|_{H^2(0,1)}. \tag{1.15}$$

where $I^h u$ is the interpolator of u, i.e., $I^h u \in V_h$ is such that $I^h u(\mathbf{x}_i) = u(\mathbf{x}_i)$ at all nodes $\mathbf{x}_i$ of the domain discretization.

Finally, to estimate how well the finite element method with piecewise linear functions perform, we choose $v_h = I^h u$, the interpolation of u in V_h. The interpolation estimate and Céa's Lemma 1.3 yield

$$\|u - u_h\|_V \le c\frac{\beta}{\alpha}h|u|_{H^2(0,1)}. \tag{1.16}$$

Note that the such error estimate guarantees that the finite element solution *converges* to the exact solution, as the mesh gets refined. However, and that is often the case, the convergence can be hindered if $\alpha \ll 1$ or $\beta|u|_{H^2(0,1)} \gg 1$.

1.3   Alternative Methods

It is often the case that, for multiscale problems, the traditional finite element method fails. We briefly present here some methods that are used to overcome the deficiencies of classical schemes. In the descriptions that follow, consider the weak and strong forms (1.1) and (1.2). Consider a partition $\mathcal{T}_h$ of the domain Ω into finite elements (intervals in one dimension, and triangles in two dimensions), and let P_1 be the space of continuous piecewise linear functions associated with such partition.

Some important ideas that we do not explore are present in mixed and discontinuous Galerkin methods, among others. There are also several developments and extensions of the ideas presented here that we do not cover. A few important references are [16, 18, 31, 58, 59, 69, 80, 86, 117, 199].

1.3.1 Residual Free Bubbles

To deal with multiscale problems in a systematic way, the *Residual-Free Bubbles* (RFB) method was proposed in [43, 97, 106–109]. The motivation is that polynomial approximations are not able to capture small scale effects, and that suggests an enrichment procedure of the approximating space by adding *bubbles*, functions that vanish at each element boundary.

After a short formalism, the conclusion is that the bubble functions satisfy the operator at the element level, with the right-hand side being the residual due to the polynomial approximation. In practice, such problems are solved exactly or approximated.

In what follows we describe the RFB method. In general, for multiscale problems, it is possible to decompose the exact solution

$$u_{\text{sol}} = u_{\text{macro}} + u_{\text{micro}},$$

where u_{macro} describe the macroscopic behavior of the solution, and u_{micro} its microscopic behavior. In the RFB method, the decomposition is

$$u_h = u_1 + u_b, \tag{1.17}$$

where u_1 is the piecewise linear part, and the bubble part u_b captures microscopic information. Consider the *enriched space* $V_h = P_1 \oplus B$, where $P_1 \subset H_0^1(\Omega)$ is as in (1.12) or (1.14), and

$$B = \{v \in H_0^1(\Omega) \, : \, v|_K \in H_0^1(K) \text{ for all } K \in \mathscr{T}_h\} \tag{1.18}$$

is the "bubble" space. The method consists in finding $u_h \in V_h$ such that

$$a(u_h, v_h) = \langle f, v_h \rangle \quad \text{for all } v_h \in V_h.$$

From (1.17), with $u_1 \in P_1$ and $u_b \in B$, it follows that

$$a(u_1 + u_b, v_1) = \langle f, v_1 \rangle \quad \text{for all } v_1 \in P_1, \tag{1.19}$$

$$a(u_1 + u_b, v_b) = \langle f, v_b \rangle \quad \text{for all } v_b \in B. \tag{1.20}$$

The problem with (1.19), (1.20) is that it is a coupled system. We next do a *static condensation*, i.e., write u_b in terms of u_1. Taking in (1.20) test functions v_b with support within an arbitrary element K, and integrating by parts, it follows that u_b is the strong solution of the local problem

$$\mathscr{L} u_b = -\mathscr{L} u_1 + f \quad \text{in } K, \qquad u_b = 0 \quad \text{on } \partial K,$$

for all elements K. Writing $u_b = T(-\mathscr{L} u_1 + f)$, replacing in (1.19), and using the problem linearity, the final formulation yields that $u_1 \in P_1$ solves

$$a(u_1, v_1) - a(T\mathscr{L} u_1, v_1) = \langle f, v_1 \rangle - a(Tf, v_1) \quad \text{for all } v_1 \in P_1. \tag{1.21}$$

One way to interpret the above formulation is as a *parameter-free stabilized method*, see Sect. 1.3.7. It can also be interpreted as an upscaling technique, where $u_1 \in P_1$ solves the "homogenized" problem

$$a^*(u_1, v_1) = \langle f^*, v_1 \rangle \quad \text{for all } v_1 \in P_1,$$

where $a^*(u_1, v_1) = a((I - T\mathscr{L})u_1, v_1)$, and $\langle f^*, v_1 \rangle = \langle f, v_1 \rangle - a(Tf, v_1)$.

Finally, it is possible to see such formulation "almost" as a Petrov–Galerkin method (in Petrov–Galerkin formulations, the space of test functions differs from the trial space). If $\{\phi_i\}_{i=1}^N$ is a basis for P_1 and $u_1 = \sum_{i=1}^N u_i\phi_i$, then

$$\sum_{i=1}^N u_i a(\lambda_i, \phi_j) = \langle f, \phi_j \rangle - a(Tf, \phi_j) \quad \text{for } j = 1, \dots, N, \tag{1.22}$$

where $\lambda_i = (I - T\mathscr{L})\phi_i$, i.e.,

$$\mathscr{L} \lambda_i = 0 \quad \text{em } K, \qquad \lambda_i = \phi_i \quad \text{on } \partial K, \tag{1.23}$$

for all $K \in \mathscr{T}_h$. The basis functions in the admissible space solve the operator locally, but the test functions remain the classical piecewise linear. In terms of implementation, one first compute all the multiscale functions λ_i solving (1.23) to then assemble the matrix entries $a(\lambda_i, \phi_j)$, and finally solve (1.22).

A nice feature of the RFB method is that $u_h(\mathbf{x}_i) = u_1(\mathbf{x}_i)$, for every node $\mathbf{x}_i$, since u_b vanishes over the edges and, in particular, at the nodal points. That means that the method requires no post-processing. Also, for one-dimensional problems, $u_h = u$, that is, the RFB method *yields the exact solution* since $H_0^1(I) = P_1 \oplus B$, if $I \subset \mathbb{R}$ is an interval. An immediate consequence is that $u_h(\mathbf{x}_i) = u(\mathbf{x}_i)$, and u_1 is pointwise exact at the nodes. This property holds only in one dimension, as the decomposition of $H_0^1(\Omega)$ as a direct sum of linear functions plus bubbles does not hold in higher dimensions.

Note that, in terms of computational costs, system (1.21) has the same size as the traditional Galerkin formulation, and is posed on the same space; only the bilinear form itself has been modified.

Now, there is of course the question of how to approximate the multiscale functions λ_i, since this requires solving equations with the same original operator. Note, however, that these problems are now local, and thus of smaller sizes. Since they are independent, parallel computations are straightforward. Another important point to make is that what matters is not so much the details of local solution, but its integral. So, maybe a rough approximation of the bubble suffices, and that can be computed or "guessed" analytically, using, for instance, asymptotic properties of the method. Also, unrefined local meshes can be employed [43, 99, 102, 103, 107, 188].

1.3.2 Multiscale Finite Element Method

Another way to capture multiscale aspects of the solution is to require that basis functions are local solutions to the original problem—the idea goes back to Trefftz [119]. So, for instance, given a partition $\mathcal{T}_h$ of Ω, we can define the basis functions space $V_h = \text{span}\{\lambda_i : i \text{ is a node of } \mathcal{T}_h\}$, with λ_i defined by (1.23). The *Multiscale Finite Element Method* (MsFEM) searches then for $u_h \in V_h$ such that

$$a(u_h, v_h) = \langle f, v_h \rangle \quad \text{for all } v_h \in V_h. \tag{1.24}$$

The performance of the method for multiscale problems is analyzed on a case-by-case basis. Note, however, the similarities with the RFB scheme. Indeed, the left-hand side of both (1.22) and (1.24) is equal, since $a(\lambda_i, \lambda_j) = a(\lambda_i, \phi_j)$. That identity follows from $a(\lambda_i, \lambda_j) = a(\lambda_i, \phi_j) - a(\lambda_i, T\mathcal{L}\phi_j)$ and since $T\mathcal{L}\phi_j \in B$, we can integrate by parts and

$$a(\lambda_i, T\mathcal{L}\phi_j) = \sum_{K \in \mathcal{T}_h} \int_K \mathcal{L}\lambda_i T\mathcal{L}\phi_j \, d\mathbf{x} = 0$$

from (1.23).

Although the RFB and MsFEM yields the same left hand sides, the methods are not equal as the right-hand sides differ; see [135] for a justification to disregard the term Tf present in the RFB. Since the subspace of the MsFEM is contained in the subspace of the RFB method, the error estimate (1.10) states that the RFB solution is at least as good as the MsFEM solution, in the energy norm.

For one-dimensional problems, the MsFEM is nodally exact. That is a characteristic of finite element methods that use basis functions that are local solutions of the very PDE they are approximating. Indeed, from an integration by parts,

$$a(u, v) = \sum_{k=0}^{N-1} \int_{x_k}^{x_{k+1}} u\mathcal{L}v \, dx + \sum_{k=0}^{N} u(x_k)Bv(x_k) \quad \text{for all } v \in V_h,$$

where B is an operator acting on v—its precise definition does not actually matter now. Then

$$a\left(u - \sum_{i=0}^{N} u(x_i)\lambda_i, \lambda_j\right) = \sum_{k=0}^{N} \int_{x_k}^{x_{k+1}} \left(u - \sum_{i=0}^{N} u(x_i)\lambda_i\right) \mathscr{L}\,\lambda_j\,dx$$

$$+ \sum_{k=0}^{N} \left(u(x_k) - \sum_{i=0}^{N} u(x_i)\lambda_i(x_k)\right) Bv(x_k) = 0,$$

where we used that $\mathscr{L}\,\lambda_j = 0$, and that $\lambda_i(x_k) = \delta_{i,k}$. The conclusion is that $u_h = \sum_{i=0}^{N} u(x_i)\lambda_i$, and thus $u_h(x_i) = u(x_i)$ for all nodes. Unfortunately, in higher dimensions, this property is lost.

Remark 1.2 As a simple particular instance, consider $a(u, v) = \int_0^1 u'v'\,dx$. Then the MsFEM recovers the classical finite element method with linear piecewise polynomials, and, as above, the discrete solution is the interpolation of the exact solution. Then, the piecewise linear finite element method is nodally exact for this problem. Note that while the RFB method yields the exact solution in the whole domain, in general the MsFEM is exact only at the nodes. To see that, consider the problem $u'' = 1$. Then the solution is parabolic, but the approximate function is piecewise linear. The details are left to the reader.

Applications of the MsFEM include equations posed on domains with rough boundaries [156] as in Chap. 5, and flows in heterogeneous media [82, 83, 87, 88, 134] as in Chap. 6.

1.3.3   Variational Multiscale Method and Localization

Using a quite broad framework, the Variational Multiscale (VMS) method tries to give a general form to derive multiscale method, based on decomposition of spaces [141, 142, 144]. The basic idea is no different from what is considered in the Residual Free Bubble method; see Sect. 1.3.1, and also [40].

However, in its full generality, the decompositions give rise to nonlocal problems. Related to that, and using very special decompositions, the Localized Orthogonal Decomposition (LOD) method is able to deal with multiscale problems assuming minimal regularity of the solutions [89, 161, 162].

Consider (1.1) and assume that $a(\cdot, \cdot)$ is symmetric, and coercive, as in (1.7). Let $V = H_0^1(\Omega)$ and $f \in L^2(\Omega)$. Consider also the Galerkin method (1.9), with $V_h = P_1$ as in (1.12) or (1.14). It follows from the Galerkin orthogonality that

$$\|u - u_h\|_{H^1(\Omega)}^2 \le \frac{1}{\alpha} a(u - u_h, u - u_h)$$

$$= \frac{1}{\alpha} a(u - u_h, u) = \langle f, u - u_h \rangle$$

$$\leq \frac{1}{\alpha} \|f\|_{L^2(\Omega)} \|u - u_h\|_{L^2(\Omega)}. \tag{1.25}$$

Consider a bounded linear operator $\mathscr{I}_h : V \to V_h$ such that $\mathscr{I}_h v_h \neq 0$ for all nonzero function $v_h \in V_h$, and

$$\|v - \mathscr{I}_h v\|_{L^2(\Omega)} \leq ch\|v\|_{H^1(\Omega)}.$$

The Clément interpolation satisfies such property [92, 162]. If

$$\mathscr{I}_h(u - u_h) = 0, \tag{1.26}$$

it follows from (1.25) that

$$\|u - u_h\|_{H^1(\Omega)}^2 \leq \frac{1}{\alpha} \|f\|_{L^2(\Omega)} \|u - u_h - \mathscr{I}_h(u - u_h)\|_{L^2(\Omega)}$$

$$\leq c \frac{h}{\alpha} \|f\|_{L^2(\Omega)} \|u - u_h\|_{H^1(\Omega)},$$

and then

$$\|u - u_h\|_{H^1(\Omega)} \leq c \frac{h}{\alpha} \|f\|_{L^2(\Omega)}. \tag{1.27}$$

Note that the estimate requires only that $f \in L^2$, and it is not necessary that $u \in H^2(\Omega)$, cf. (1.16). The problem is that (1.26) was fundamental to derive the estimate. The next step is to construct a method that guarantees it.

Consider

$$V^{\mathrm{f}} = \ker \mathscr{I}_h = \{v \in H_0^1(\Omega) : \mathscr{I}_h v = 0\},$$

assume that its codimension is dim V_h, and let the fine scale projector $\mathscr{F} : V \to V^{\mathrm{f}}$ be defined by

$$a(\mathscr{F} v, v^{\mathrm{f}}) = a(v, v^{\mathrm{f}}) \quad \text{for all } v^{\mathrm{f}} \in V^{\mathrm{f}}. \tag{1.28}$$

Consider also the space

$$V_h^{\mathrm{ms}} = \{\mathscr{F} v_h - v_h : v_h \in V_h\}.$$

Now, in the spirit of the VMS method we can decompose V as a direct sum of a *multiscale space* (where we want to compute the solutions) and a *fine scale space*, that is $V = V_h^{\mathrm{ms}} \oplus V^{\mathrm{f}}$. To see this, note first that $a(v_h^{\mathrm{ms}}, v^{\mathrm{f}}) = 0$ for all $v_h^{\mathrm{ms}} \in V_h^{\mathrm{ms}}$ and

$v^f \in V^f$. Then, for an arbitrary $v \in V$ there is a unique decomposition $v = v_h^{ms} + v^f$, where $v_h^{ms} \in V_h^{ms}$ and $v^f \in V^f$. Indeed, v^f solves $a(v^f, w^f) = a(v, w^f)$ for all $w^f \in V^f$, and $v_h^{ms} = v - v^f$.

We define now the Galerkin solution by solving

$$a(u_h^{ms}, v_h^{ms}) = \langle f, v_h^{ms} \rangle \quad \text{for all } v_h^{ms} \in V_h^{ms}. \tag{1.29}$$

Since now the Galerkin orthogonality $a(u - u_h^{ms}, v_h^{ms}) = 0$ holds for all $v_h^{ms} \in V_h^{ms}$, it follows that $u - u_h^{ms} \in V^f$, and then (1.26) holds. Thus the method defined by (1.29) attains the convergence rate (1.27), no matter how rough the coefficients are.

There is a caveat though. Building a basis for the multiscale space V_h^{ms} is hard: given a basis $\{\lambda_i\}_{i=1}^N$ for V_h, it is necessary to find ϕ_i such that

$$a(\phi_i, v^f) = a(\lambda_i, v^f) \quad \text{for all } v^f \in V^f, \tag{1.30}$$

and then $V_h^{ms} = \text{span}\{\lambda_i - \phi_i\}$. But solving such problem is as hard as solving the original equation. The situation is even more dismaying since (1.30) is a global problem and yield global solutions. Since the basis functions no longer have local support, a keystone property of the finite element method is lost. That renders the method, as is, useless.

It turns out that although the ϕ_i have global support, it decays exponentially fast away from the support of λ_i, at least for the problem described at Example 1.2. It is possible then to solve (1.30) locally, in a domain slightly greater than the support of λ_i, and use the resulting functions as a basis of a now modified V_h^{ms}.

1.3.4 Heterogeneous Multiscale Method

We briefly describe here the *Heterogeneous Multiscale Method* (HMM) [73–79, 168], considering once again the problem (1.5). If there is an effective matrix A that incorporates the microscale effects, the bilinear form

$$\int_\Omega (A \nabla V) \cdot \nabla W \, d\mathbf{x} = \sum_{K \in \mathscr{T}_h} \int_K (A \nabla V) \cdot \nabla W \, d\mathbf{x} \quad \text{for } V, W \in P_1,$$

would be adequate to seek an approximation for the original solution. Considering an element $K \in \mathscr{T}_h$ and the quadrature

$$\int_K p(x) \, d\mathbf{x} \approx \sum_{l=1}^L w_l p(x_l)$$

where w_l are weights and x_l are the quadrature points, we have that

$$\int_K (A \nabla V) \cdot \nabla W \, d\mathbf{x} \approx \sum_{l=1}^L w_l [(A \nabla V) \cdot \nabla W](x_l).$$

Approximate next $[(A \nabla V) \cdot \nabla W](x_l)$ in the following way. Consider $I_\delta(x_l)$ the square of size δ centered in x_l. Now, given $V \in P_1$ find $v_l = R(V)$ such that

$$-\operatorname{div}[a^\epsilon(x) \nabla v_l(x)] = 0 \quad \text{in } I_\delta(x_l),$$

$$v_l = V \quad \text{on } \partial I_\delta(x_l).$$

Consider then

$$[(A \nabla V) \cdot \nabla W](x_l) \approx \frac{1}{\delta} \int_{I_\delta(x_l)} [a^\epsilon(x) \nabla v_l(x)] \cdot \nabla w_l(x)\, d\mathbf{x},$$

where $v_l = R(V)$ and $w_l = R(W)$.

Remark 1.3 The choice of δ depends on the problem being considered. For instance, for periodic problems, δ can be the period itself. The boundary conditions used to define the operator $R(\cdot)$ can also be changed, for instance considering $V - R(V)$ periodic in $I_\delta(x_l)$.

In the periodic case, the approximation error is given by

$$\|U - U_{\text{HMM}}\|_{H^1(\Omega)} \le C(h + \epsilon),$$

but that is achieved if the definition of the cell problem is carefully geared towards solving the very specific periodic case problem.

1.3.5 Hybrid Methods

A powerful way to deal with multiscale problems is using *hybrid methods* [176]. Hybrid schemes have a lot in common with domain decomposition techniques [114, 192], in the sense that they "break" an initially coupled problem into a set of independent problems that are latter "glued" together [17, 31, 111, 115, 192]. A mathematical presentation and analysis of the method can be found in the outstanding paper [180]. The basic idea of hybrid methods originated several powerful schemes that rely on parallel computation to solve problems that can be of multiscale nature, among then the *Finite Element Tearing and Interconnecting* (FETI) method [95], the *Total FETI* (TFETI) [71, 151], the *Discontinuous Enrichment Method* (DEM) [94], the *Hybridizable Discontinuous Galerkin* (HDG) method [58, 59], and the *Multiscale Hybrid-Mixed* (MHM) method [15, 122–124, 157].

Considering again (1.5), we derive a primal hybrid variational formulation that differs from (1.6). We start by considering a partition of $(0, 1)$ into the elements (x_i, x_{i+1}), for $i = 0, \ldots, N$, as in Sect. 1.2. Multiplying (1.5) by a function $v \in H^1(x_i, x_{i+1})$ and integrating by parts in the corresponding interval yields

$$\int_{x_i}^{x_{i+1}} a^\epsilon \frac{du^\epsilon}{dx} \frac{dv}{dx}\, dx + \frac{d(a^\epsilon u^\epsilon)}{dx}(x_{i+1}) v(x_{i+1}) - \frac{d(a^\epsilon u^\epsilon)}{dx}(x_i) v(x_i) = \int_{x_i}^{x_{i+1}} f(x) v(x)\, dx.$$

To end up with a global formulation, we introduce the space

$$H^1(\mathcal{T}_h) = \{v \in L^2(0, 1) : v|_{(x_i, x_{i+1})} \in H^1(x_i, x_{i+1}), \ i = 0, \dots, N\}.$$

Note that the functions in the above space are not necessarily continuous at the nodes, only *piecewise continuous*.

Denoting $\lambda_i = d(a^\epsilon u^\epsilon)/dx(x_i)$, and adding over all elements, we gather that

$$\sum_{i=0}^{N} \int_{x_i}^{x_{i+1}} a^\epsilon \frac{du^\epsilon}{dx} \frac{dv}{dx} \, dx + \sum_{i=0}^{N+1} \lambda_i \llbracket v \rrbracket (x_i) = \int_0^1 f(x)v(x) \, dx, \qquad (1.31)$$

where we define the jump operator

$$\llbracket v \rrbracket (x_i) = \lim_{\substack{h \to 0 \\ h > 0}} v(x_i - h) - v(x_i + h) \quad \text{for } i = 1, \dots, N,$$

$$\llbracket v \rrbracket (x_0) = -v(x_0), \qquad \llbracket v \rrbracket (x_{N+1}) = v(x_{N+1}).$$

Consider the notation

$$\langle \boldsymbol{\xi}, \llbracket v \rrbracket \rangle = \sum_{i=0}^{N+1} \xi_i \llbracket v \rrbracket (x_i).$$

If $v \in H_0^1(0, 1)$, and therefore is continuous, then $\llbracket v \rrbracket (x_i) = 0$ at all nodes. Conversely, if $\langle \boldsymbol{\xi}, \llbracket v \rrbracket \rangle = 0$ for all $\boldsymbol{\xi} \in \mathbb{R}^{N+2}$, then $v \in H_0^1(0, 1)$ and is continuous.

Now, if a smooth u^ϵ solves (1.31), is continuous and satisfies the Dirichlet boundary conditions, then it also solves (1.5). However, if we are to search a solution within $H^1(\mathcal{T}_h)$, we actually need to enforce its continuity at the interior nodes, as well as the boundary conditions. That can be done by requiring that $\llbracket u^\epsilon \rrbracket \xi_i = 0$ for all $\xi_i \in \mathbb{R}$ and $i = 0, \dots, N + 1$. We conclude that $u^\epsilon \in H^1(\mathcal{T}_h)$, $\boldsymbol{\lambda} \in \mathbb{R}^{N+2}$ solve

$$\sum_{i=0}^{N} \int_{x_i}^{x_{i+1}} a^\epsilon \frac{du^\epsilon}{dx} \frac{dv}{dx} \, dx + \langle \boldsymbol{\lambda}, \llbracket v \rrbracket \rangle = \int_0^1 f(x)v(x) \, dx \quad \text{for all } v \in H^1(\mathcal{T}_h),$$

$$\langle \boldsymbol{\xi}, \llbracket u^\epsilon \rrbracket \rangle = 0 \quad \text{for all } \boldsymbol{\xi} \in \mathbb{R}^{N+2}. \qquad (1.32)$$

The above system is in *mixed form* [31, 111], and a full account of the area is beyond our scope. Note, however, that (1.32) is well posed. Indeed, it is possible to construct a solution starting from (1.6). Also, any solution of (1.32) must also solve (1.6): it must be continuous from the second equation of (1.32), and then it is enough to choose $v \in H_0^1(0, 1)$ in the first equation. Then uniqueness follows.

So, we derived another formulation of the original problem. What is interesting in this time is that (1.32) can be solved efficiently, by a *localization* procedure. Fix $i \in \{0, \dots, N\}$. By considering arbitrary $v \in H^1(\mathcal{T}_h)$ vanishing outside (x_i, x_{i+1}), we see that u^ϵ solve a second order problem with Neumann boundary condition. This kind of problem does not have unique solution in general—it is defined up to a constant. That justifies splitting the solution into a constant plus its average. Consider the spaces

$$\mathscr{C} = \{v \in H^1(\mathscr{T}_h) : v|_{(x_i,x_{i+1})} \text{ is constant, } i = 0, \ldots, N\}.$$

$$\mathring{H}^1(\mathscr{T}_h) = \{v \in H^1(\mathscr{T}_h) : \int_{x_i}^{x_{i+1}} v \, dx = 0, \ i = 0, \ldots, N\},$$

Then

$$H^1(\mathscr{T}_h) = \mathscr{C} \oplus \mathring{H}^1(\mathscr{T}_h), \tag{1.33}$$

i.e., for any $v \in H^1(\mathscr{T}_h)$, there exist unique $\bar{v} \in \mathscr{C}$ and $\mathring{v} \in \mathring{H}^1(\mathscr{T}_h)$ such that $v = \bar{v} + \mathring{v}$. From the definition, $\bar{v}|_{(x_i,x_{i+1})} = \int_{x_i}^{x_{i+1}} v \, dx/(x_{i+1} - x_i)$, and $\mathring{v} = v - \bar{v}$.

We now rewrite (1.32) by splitting $u^\epsilon = \bar{u}^\epsilon + \mathring{u}^\epsilon$, where $\bar{u}^\epsilon \in \mathscr{C}$, $\mathring{u}^\epsilon \in \mathring{H}^1(\mathscr{T}_h)$ and $\boldsymbol{\lambda} \in \mathbb{R}^{N+2}$ solve

$$\langle \boldsymbol{\lambda}, \llbracket \bar{v} \rrbracket \rangle = \int_0^1 f\bar{v} \, dx \quad \text{for all } \bar{v} \in \mathscr{C},$$

$$\sum_{i=0}^N \int_{x_i}^{x_{i+1}} a^\epsilon \frac{d\mathring{u}^\epsilon}{dx} \frac{d\mathring{v}}{dx} \, dx + \langle \boldsymbol{\lambda}, \llbracket \mathring{v} \rrbracket \rangle = \int_0^1 f\mathring{v} \, dx \quad \text{for all } \mathring{v} \in \mathring{H}^1(\mathscr{T}_h), \tag{1.34}$$

$$\langle \boldsymbol{\xi}, \llbracket \bar{u}^\epsilon + \mathring{u}^\epsilon \rrbracket \rangle = 0 \quad \text{for all } \boldsymbol{\xi} \in \mathbb{R}^{N+2}.$$

We next do a static condensation to write $\mathring{u}^\epsilon$ in terms of $\boldsymbol{\lambda}$ and f. Note that we can write the second equation in (1.34) as a set of local problems, by considering test functions that vanish everywhere except at a single element. Consider $T\boldsymbol{\lambda} \in \mathring{H}^1(\mathscr{T}_h)$ and $\hat{T}f \in \mathring{H}^1(\mathscr{T}_h)$ solving

$$\int_{x_i}^{x_{i+1}} a^\epsilon \frac{d\mathring{u}^\epsilon}{dx} \frac{d\mathring{v}}{dx} \, dx = \lambda_i \mathring{v}(x_i) - \lambda_{i+1}\mathring{v}(x_{i+1}), \qquad \int_{x_i}^{x_{i+1}} a^\epsilon \frac{d\mathring{u}^\epsilon}{dx} \frac{d\mathring{v}}{dx} \, dx = \int_0^1 f\mathring{v} \, dx, \tag{1.35}$$

for all $\mathring{v} \in \mathring{H}^1(x_i, x_{i+1})$ and $i = 0, \ldots, N$. Here, $\mathring{H}^1(x_i, x_{i+1})$ is the set of zero average functions in $H^1(x_i, x_{i+1})$.

Remark 1.4 It is left to the reader to prove that the problems in (1.35) are well posed, and to rewrite it in its strong form.

We can now rewrite (1.34) by considering that $\mathring{u}^\epsilon = T\boldsymbol{\lambda} + \hat{T}f$ implies that $\bar{u}^\epsilon \in \mathscr{C}$ and $\boldsymbol{\lambda} \in \mathbb{R}^{N+2}$ solve

$$\langle \boldsymbol{\xi}, \llbracket T\boldsymbol{\lambda} \rrbracket \rangle + \langle \boldsymbol{\xi}, \llbracket \bar{u}^\epsilon \rrbracket \rangle = -\langle \boldsymbol{\xi}, \llbracket \hat{T}f \rrbracket \rangle \quad \text{for all } \boldsymbol{\xi} \in \mathbb{R}^{N+2},$$

$$\langle \boldsymbol{\lambda}, \llbracket \bar{v} \rrbracket \rangle = \int_0^1 f\bar{v} \, dx \quad \text{for all } \bar{v} \in \mathscr{C}. \tag{1.36}$$

After solving the system, it is possible to recover the original solution from the formula

$$u^\epsilon = \bar{u}^\epsilon + T\lambda + \hat{T}f. \tag{1.37}$$

Note that the (1.36) is in the mixed form, and is finite dimensional, since $\mathscr{C}$ is of finite dimension. So, to prove existence and uniqueness of solutions, it is enough to show uniqueness. That follows, for instance, from an argument by contradiction and the uniqueness of the original problem.

Observe now that $\lambda = \sum_{i=0}^{N+1} \lambda_i \mathbf{e}_i$, where the $\mathbf{e}_i$ are the vectors of the canonical basis of $\mathbb{R}^{N+2}$. Then, if $\boldsymbol{\psi}_i = [\![T\mathbf{e}_i]\!]$, we gather that $[\![T\lambda]\!] = \sum_{i=0}^{N+1} \lambda_i \boldsymbol{\psi}_i$. By writing $\bar{u}^\epsilon = \sum_{i=0}^{N} \bar{u}_i \chi_i$, where χ_i is the characteristic function of the set (x_i, x_{i+1}), we have that (1.36) is equivalent to

$$\sum_{i=0}^{N+1} \lambda_i \langle \mathbf{e}_j, \boldsymbol{\psi}_i \rangle + \sum_{i=0}^{N} \bar{u}_i \langle \mathbf{e}_j, [\![\chi_i]\!] \rangle = -\langle \mathbf{e}_j, [\![\hat{T}f]\!] \rangle \quad \text{for } j = 0, \dots, N+1,$$

$$\sum_{i=0}^{N+1} \lambda_i \langle \boldsymbol{\psi}_i, [\![\chi_j]\!] \rangle = \int_0^1 f\chi_j \, dx \quad \text{for } j = 0, \dots, N. \tag{1.38}$$

The reason why (1.38) is efficient is the following. The problem under consideration is of multiscale type, but this is not apparent in (1.38). The only multiscale problem that has to be solved is to compute $\boldsymbol{\psi}_i$ and $\hat{T}f$, and that can be done in parallel. Moreover, these problems are local and thus much easier to solve than the original one. This features resembles that of RFB and MsFEM, but now the local problems are defined by Neumann instead of Dirichlet boundary conditions.

All above arguments and manipulations were performed at the continuous level, and therefore yields the exact solution. Of course, in practice the operators T and $\hat{T}$ have to be discretized, at least in higher dimensions. That introduces an error in the system that needs to be taken into account in the error analysis.

It might come as a surprise that, although all above computations become simpler in the one-dimensional case, its generalization to higher dimensional situations is remarkably similar, the major difference being that the theoretical justification of the method becomes harder. In particular, as the nodes become edges and faces, the space for λ is no longer finite dimensional, but a delicate space $H^{-1/2}$, that requires care [111, 113, 164].

1.3.6 Extending the Discontinuous Enriched Method

System (1.36) yields $\bar{u}^\epsilon$ as an approximation of u^ϵ, and that is the elementwise average of the exact solution. This might be considered poor for some applications. Of course, the post-processing (1.37) cures that, but this means in practice that the discretized versions of the multiscale operators T and $\hat{T}$ have to be stored in memory. For multiscale problems, that might be an expensive request.

We want to develop a method that can deliver higher order approximations without resorting to post-processing, but still keeping the nice properties of hybrid methods. Expanding some ideas present in the *Discontinuous Enrichment Method* (DEM) [94], this is possible [157].

Consider $\overset{\circ}{V}_p \subset H_0^1(0,1) \cap \overset{\circ}{H}^1(\mathcal{T}_h)$ the space of continuous piecewise polynomials of degree p that has zero average in each element, and let $\overset{\circ}{H}^1(\mathcal{T}_h) = \overset{\circ}{V}_p \oplus \overset{\circ}{V}_e$, for some $\overset{\circ}{V}_e \subset \overset{\circ}{H}^1(\mathcal{T}_h)$ (the space of discontinuous functions $\overset{\circ}{V}_e$ could be, for instance, the orthogonal complement of $\overset{\circ}{V}_p$ with respect to the inner product $a(\cdot, \cdot)$). Consider then (cf. (1.33)) the three-field decomposition

$$H^1(\mathcal{T}_h) = \mathscr{C} \oplus \overset{\circ}{V}_p \oplus \overset{\circ}{V}_e.$$

Proceeding as in Sect. 1.3.5, we let $u^\epsilon = \bar{u}^\epsilon + \overset{\circ}{u}_p^\epsilon + \overset{\circ}{u}^\epsilon$, where $\bar{u}^\epsilon \in \mathscr{C}$, $\overset{\circ}{u}_p^\epsilon \in \overset{\circ}{V}_p$, $\overset{\circ}{u}_e^\epsilon \in \overset{\circ}{V}_e$ and $\boldsymbol{\lambda} \in \mathbb{R}^{N+2}$ solve (cf. (1.34))

$$\langle \boldsymbol{\lambda}, [\![\bar{v}]\!]\rangle = \int_0^1 f\bar{v}\,dx \quad \text{for all } \bar{v} \in \mathscr{C},$$

$$\int_0^1 a^\epsilon \frac{d\overset{\circ}{u}_p^\epsilon}{dx}\frac{d\overset{\circ}{v}_p}{dx}\,dx + \sum_{i=0}^N \int_{x_i}^{x_{i+1}} a^\epsilon \frac{d\overset{\circ}{u}_e^\epsilon}{dx}\frac{d\overset{\circ}{v}_p}{dx}\,dx = \int_0^1 f\overset{\circ}{v}_p\,dx \quad \text{for all } \overset{\circ}{v}_p \in \overset{\circ}{V}_p,$$

$$\sum_{i=0}^N \int_{x_i}^{x_{i+1}} a^\epsilon \frac{d\overset{\circ}{u}_p^\epsilon}{dx}\frac{d\overset{\circ}{v}_e}{dx}\,dx + \sum_{i=0}^N \int_{x_i}^{x_{i+1}} a^\epsilon \frac{d\overset{\circ}{u}_e^\epsilon}{dx}\frac{d\overset{\circ}{v}_e}{dx}\,dx + \langle \boldsymbol{\lambda}, [\![\overset{\circ}{v}_e]\!]\rangle$$

$$= \int_0^1 f\overset{\circ}{v}_e\,dx \quad \text{for all } \overset{\circ}{v}_e \in \overset{\circ}{V}_e,$$

$$\langle \boldsymbol{\xi}, [\![\bar{u}^\epsilon + \overset{\circ}{u}_e^\epsilon]\!]\rangle = 0 \quad \text{for all } \boldsymbol{\xi} \in \mathbb{R}^{N+2}. \tag{1.39}$$

Note that we use above that $\langle \boldsymbol{\lambda}, [\![\overset{\circ}{v}_p]\!]\rangle = \langle \boldsymbol{\xi}, [\![\overset{\circ}{u}_p^\epsilon]\!]\rangle = 0$, since the functions in $\overset{\circ}{V}_p$ are continuous.

Let $T_e^p : \overset{\circ}{V}_p \to \overset{\circ}{V}_e$, $T_e : \mathbb{R}^{N+2} \to \overset{\circ}{V}_e$, and $\hat{T}_e : L^2(0,1) \to \overset{\circ}{V}_e$ be such that (cf. (1.35))

$$\int_{x_i}^{x_{i+1}} a^\epsilon \frac{dT_e^p\overset{\circ}{u}_e^\epsilon}{dx}\frac{d\overset{\circ}{v}_e}{dx}\,dx = -\int_{x_i}^{x_{i+1}} a^\epsilon \frac{d\overset{\circ}{u}_p^\epsilon}{dx}\frac{d\overset{\circ}{v}_e}{dx}\,dx,$$

$$\int_{x_i}^{x_{i+1}} a^\epsilon \frac{T_e\boldsymbol{\lambda}}{dx}\frac{d\overset{\circ}{v}_e}{dx}\,dx = \lambda_i\overset{\circ}{v}_e(x_i) - \lambda_{i+1}\overset{\circ}{v}_e(x_{i+1}), \tag{1.40}$$

$$\int_{x_i}^{x_{i+1}} a^\epsilon \frac{d\hat{T}_e f}{dx}\frac{d\overset{\circ}{v}_e}{dx}\,dx = \int_{x_i}^{x_{i+1}} f\overset{\circ}{v}_e\,dx.$$

for all $\overset{\circ}{v}_e \in \overset{\circ}{V}_e$ and $i = 0, \ldots, N$. It follows from the third equation of (1.39) that

$$\overset{\circ}{u}_e^\epsilon = T_e^p\overset{\circ}{u}_p^\epsilon + T_e\boldsymbol{\lambda} + \hat{T}_e f. \tag{1.41}$$

Thus, from (1.39),

$$\langle \boldsymbol{\xi}, [\![T_e \boldsymbol{\lambda}]\!] \rangle + \langle \boldsymbol{\xi}, [\![T_e^p \mathring{u}_p^\epsilon]\!] \rangle + \langle \boldsymbol{\xi}, [\![\bar{u}^\epsilon]\!] \rangle = -\langle \boldsymbol{\xi}, [\![\hat{T}_e f]\!] \rangle$$

$$\sum_{i=0}^{N} \int_{x_i}^{x_{i+1}} a^\epsilon \frac{dT_e \boldsymbol{\lambda}}{dx} \frac{d\mathring{v}_p}{dx} \, dx + \sum_{i=0}^{N} \int_{x_i}^{x_{i+1}} a^\epsilon \frac{d(\mathring{u}_p^\epsilon + T_e^p \mathring{u}_p^\epsilon)}{dx} \frac{d\mathring{v}_p}{dx} \, dx$$

$$= \int_0^1 f \mathring{v}_p \, dx - \sum_{i=0}^{N} \int_{x_i}^{x_{i+1}} a^\epsilon \frac{d\hat{T}_e f}{dx} \frac{d\mathring{v}_p}{dx} \, dx$$

$$\langle \boldsymbol{\lambda}, [\![\bar{v}]\!] \rangle = \int_0^1 f \bar{v} \, dx \tag{1.42}$$

for all $\boldsymbol{\xi} \in \mathbb{R}^{N+2}$, $\mathring{v}_p \in \mathring{V}_p$, and $\bar{v} \in \mathscr{C}$.

The method approximates the original solution by $\bar{u}^\epsilon + \mathring{u}_p^\epsilon$, but the extra term $\mathring{u}_e^\epsilon$ can be added if needed. In terms of implementation, $\mathring{V}_p$ is finite dimensional, so no discretization is necessary, even in higher dimensions. The method requires computing the operators T_e, T_e^p, and $\hat{T}_e$, and the local problems (1.40) are equivalent to (1.35). The size of the system (1.42) is larger than (1.38), increased by the dimension of $\mathring{V}_p$, and the system is still of mixed form. Finally, note the savings with respect to computer memory, since it is no longer necessary to store the solutions of problems (1.40) for post-processing. Indeed, after these solutions are obtained, they can be used to assemble the matrices in (1.42) and then discarded.

1.3.7   Stabilized Methods

There are of course several other methods in the literature that try to overcome potential deficiencies of the Galerkin method. Among them, stabilized methods were developed mainly in the 1970s and 1980s, but are still very popular. The design of the method was initially motivated to overcome the possible lack of stability of advection-diffusion equations [44, 139, 140].

Consider the one-dimensional problem

$$-\varepsilon \frac{d^2 u^\epsilon}{dx^2} + \frac{du^\epsilon}{dx} = f, \quad \text{in } (0, 1), \qquad u^\epsilon(0) = 1, \quad u^\epsilon(1) = 0,$$

where ε is a small positive real number. Then the Galerkin solution fails to deliver reasonable solutions; see Chap. 2, for instance. In the weak formulation, $u^\epsilon \in H^1(0, 1)$ is such that $u^\epsilon(0) = 1$, $u^\epsilon(1) = 0$ and $a(u^\epsilon, v) = \langle f, v \rangle$ for all $v \in H_0^1(0, 1)$, where

$$a(u, v) = \int_0^1 \varepsilon \frac{du}{dx} \frac{dv}{dx} + \frac{du}{dx} v \, dx, \qquad \langle f, v \rangle = \int_0^1 f v \, dx.$$

In general, the basic idea behind the method is to increase stability (coercivity) while keeping good properties of the finite element method, and does so by modifying the bilinear form $a(\cdot, \cdot)$. A possible stabilization scheme seeks $u_1 \in P_1$ such that

$$a(u_1, v_1) + S(f, u_1, v_1) = \langle f, v_1 \rangle \quad \text{for all } v_1 \in P_1. \tag{1.43}$$

Here, P_1 is again the space of continuous piecewise linear polynomials.

An important concept in finite element analysis is *consistency*; a method is consistent if the exact solution also satisfies the discrete formulation. For stabilized methods, consistency is preserved if $S(f, u, v_1) = 0$. Designing S is an artistic endeavor, and usually justified a posteriori through error analyses. Some stabilizing choices are

$$S(f, u_1, v_1) = \sum_{i=0}^{N} \tau_i \begin{cases} \int_{x_i}^{x_{i+1}} (-\varepsilon u_1'' + u_1' - f) v_1' \, dx, \\ \int_{x_i}^{x_{i+1}} (-\varepsilon u_1'' + u_1' - f)(-\varepsilon v_1'' + v_1') \, dx, \\ \int_{x_i}^{x_{i+1}} (-\varepsilon u_1'' + u_1' - f)(\varepsilon v_1'' + v_1') \, dx, \end{cases}$$

where τ_i are *stabilizing parameters*, to be defined [44, 96, 98, 100, 110, 121, 126, 139, 140, 143, 146].

Note that consistency is preserved since $S(f, u, v_1) = 0$, and that the method is *conforming*. Note also that, for piecewise linear elements, all the above methods coincide. Actually, at its origin, stabilized methods were restricted to linear elements and the design of parameters was inspired by optimal finite difference schemes.

There are wide range of choices for τ_i, and a particular one for the linear case is

$$\tau_i = \frac{h}{2} \begin{cases} \mathrm{Pe}_i & \text{if } 0 \leq \mathrm{Pe}_i < 1, \\ 1 & \text{if } \mathrm{Pe}_i \geq 1, \end{cases} \qquad \mathrm{Pe}_i = \frac{h}{6\varepsilon}.$$

The motivation behind this definition is that, if $\varepsilon \ll h$, then it is necessary to stabilize the problem, and $\tau = h/2$. Otherwise, if $h \ll \varepsilon$, the classical method is good enough and there is no need for extra stabilization. In this case $\tau = h^2/(12\varepsilon)$ adds only a small perturbation.

Comparing (1.21) and (1.43), it is possible to see that there is a close connection between stabilized and enriched methods, and such connection paved the way to design stabilized methods in a more consistent form [39]. The stabilization is driven by bubble functions, and there is no need to design stabilizing parameters.

Based on [61], we make such connection explicit. Consider an enrichment strategy, not necessarily the RFB method. So, as in (1.17), let $u_h = u_1 + u_b$, where u_1 is piecewise linear, and $u_b = \sum_{i=0}^{N} c_i \psi_i$, where $\psi_i \in B$ is the bubble part (see (1.18)). In the RFB method, the bubbles are determined by local problems. But other options are possible, for instance choosing quadratic polynomials of the form $\psi_i|_{(x_i, x_{i+1})} = 4(x - x_i)(x_{i+1} - x)/h^2$.

Next, we assume that f is constant within each element. Similarly to (1.20), we impose

$$a(u_1 + u_b, \psi_j) = \langle f, \psi_j \rangle, \tag{1.44}$$

for $j = 0, \ldots, N$. Then, using integration by parts, and that f and u_1' are constants over each element, we gather that

$$c_j \varepsilon \| \psi_j' \|_{L^2(x_j, x_{j+1})}^2 = (f - u_1')|_{(x_j, x_{j+1})} \int_{x_j}^{x_{j+1}} \psi_j \, dx, \tag{1.45}$$

and that determines c_j in terms of u_1 and f. Next, from (see (1.19))

$$a(u_1 + u_b, v_1) = \langle f, v_1 \rangle \quad \text{for all } v_1 \in P_1.$$

But

$$a(u_b, v_1) = \sum_{i+i}^{N} c_i a(\psi_i, v_1) = -\sum_{i+i}^{N} \tau_i (f - u_1')|_{(x_j, x_{j+1})} v_1' \int_{x_i}^{x_{i+1}} 1 \, dx = S(f, u_1, v_1),$$

where we define

$$S(f, u_1, v_1) = \sum_{i+i}^{N} \tau_i \int_{x_i}^{x_{i+1}} (-\varepsilon u_1'' + u_1' - f)(-\varepsilon v_1'' + v_1') \, dx,$$

$$\tau_i = \frac{\left(\int_{x_i}^{x_{i+1}} \psi_j \, dx \right)^2}{\varepsilon(x_{i+1} - x_i) \| \psi_j' \|_{L^2(x_j, x_{j+1})}^2}.$$

We then have the stabilized method

$$a(u_1, v_1) + S(f, u_1, v_1) = \langle f, v_1 \rangle \quad \text{for all } v_1 \in P_1.$$

The advantage of the approach just described is that there is no need to design a stabilization parameter a priori (considered to be an art for a few). The parameter naturally inherits properties from the bubble part. And, as we remarked before, the bubble can be simply computed exactly or approximately by solving local problems.

1.4 Conclusions

This chapter contains the basic definitions and notations used throughout the notes. It also contains a short account of several, but not all, finite element methods that can be used for multiscale problems. What these methods have in common is that they try to capture microscale information needed to derive a macroscale scheme.

These methods are no panacea however, and the user might have to make modifications with specific applications in mind. It is important nonetheless to understand the pros and cons of each method, allowing for educated choices.

Chapter 2
One-Dimensional Singular Perturbed Problems

Abstract In this chapter, we introduce a singular perturbed problem and a numerical difficulty associate with its discretization. We first consider the one-dimensional advective dominated advection-diffusion problem, both in terms of numerical solutions and its asymptotic expansion. We then consider a more general asymptotic expansion, including a reaction term in the equation and considering the situation when the coefficients might depend on x as well.

2.1 Introduction

One of the most important techniques in asymptotic analysis is the *matching asymptotic expansion* [62, 133, 145, 186, 195], important to understand how certain ODEs or PDEs depends on one or more parameters. Sometimes it is possible to express the solution as a *regular (or outer) expansion* that satisfies the equation, but maybe not some or all boundary conditions. In the simplest case, the expansion is a formal infinite power series in terms of the equation parameters. It is formal in the sense that we do not require the series to converge. As such expansion is being obtained, other terms are added to it to force the series to satisfy the boundary conditions. Those are the *boundary correctors* and constitute the *inner expansion*.

In what follows we work the details of these ideas considering simple examples in one-dimension, to avoid many technicalities. The differential equations have a second order (diffusion) term, and lower order terms: first order (advection), and possibly zeroth order (reaction). In the examples, the diffusion term is multiplied by a small positive scalar ε.

A main application of these problems is to understand fluids with low viscosity, in particular the Navier-Stokes system of equations (we do not consider this here). Numerically, it is not easy to design computational methods that perform well for all $\varepsilon > 0$, and equations like we consider below (specially in higher dimensions) are a perfect playground for those who want to develop and analyze robust numerical schemes [42–44, 58, 98, 100–105, 110, 121, 124, 139, 143, 146, 186].

In the next section, we consider a simple advection-diffusion equation and its finite element approximation. We then develop an asymptotic expansion of the exact solution. In Sect. 2.3 we consider a more general equation that includes non-constant advection and reaction terms, before concluding the chapter with some comments.

© The Author(s) 2017   23
A.L. Madureira, *Numerical Methods and Analysis of Multiscale Problems*,
SpringerBriefs in Mathematics, DOI 10.1007/978-3-319-50866-5_2

2.2 Advection-Diffusion with Constant Coefficients

We consider a very simple case of singular perturbed problem as well as its finite element discretization in Sect. 2.2.1. Then, in Sect. 2.2.2 we analyze what goes wrong with the classical discretization. Finally, in Sect. 2.2.3 we develop the asymptotic expansion of the solution.

2.2.1 The Problem and Its Finite Element Discretization

Consider the following boundary value problem:

$$-\varepsilon \frac{d^2 u^\epsilon}{dx^2} + \frac{du^\epsilon}{dx} = 0,$$

$$u^\epsilon(0) = 1, \quad u^\epsilon(1) = 0,$$

$$\tag{2.1}$$

where ε is a positive real number. It is convenient to assume that $\varepsilon \leq 1$. The exact solution is simply

$$u^\epsilon(x) = 1 - \frac{e^{x/\varepsilon} - 1}{e^{1/\varepsilon} - 1}. \tag{2.2}$$

The function plots for $\varepsilon = 1$, $\varepsilon = 0.1$, and $\varepsilon = 0.01$ follow in Fig. 2.1. It is clear that when ε approaches zero, there is the onset of a boundary layer close to $x = 1$. This is also highlighted by the following fact:

$$1 = \lim_{\varepsilon \to 0} \lim_{\substack{x \to 1 \\ x < 1}} u^\epsilon(x) \neq \lim_{\substack{x \to 1 \\ x < 1}} \lim_{\varepsilon \to 0} u^\epsilon(x) = 0.$$

The classification *singular perturbed* just means that we cannot impose $\varepsilon = 0$ in (2.1) and hope that $\lim_{\varepsilon \to 0} u^\epsilon$ would solve the resulting equation. Indeed, if $\varepsilon = 0$, then it follows from (2.1) a first order equation with two boundary conditions, and such problem does not have solutions in general. The limiting equation imposes that u^ϵ is a constant, but from the boundary conditions such constant has to be equal to one at $x = 0$, and to zero at $x = 1$, a clear impossibility. Note that the limit of $u^\epsilon(x)$ is the discontinuous function

$$\lim_{\varepsilon \to 0} u^\epsilon(x) = \begin{cases} 1 & \text{in } [0, 1), \\ 0 & \text{at } x = 1. \end{cases}$$

So, the pointwise limit of the exact solution satisfies the exact equation for $\varepsilon = 0$, and one of the boundary conditions. But not both.

Fig. 2.1 Exact solutions
(2.2) of the singular perturbed
problem (2.1) for $\varepsilon = 1, 0.1,$
0.01. Note the onset of a
boundary layer close to $x = 1$
as the value of ε decreases

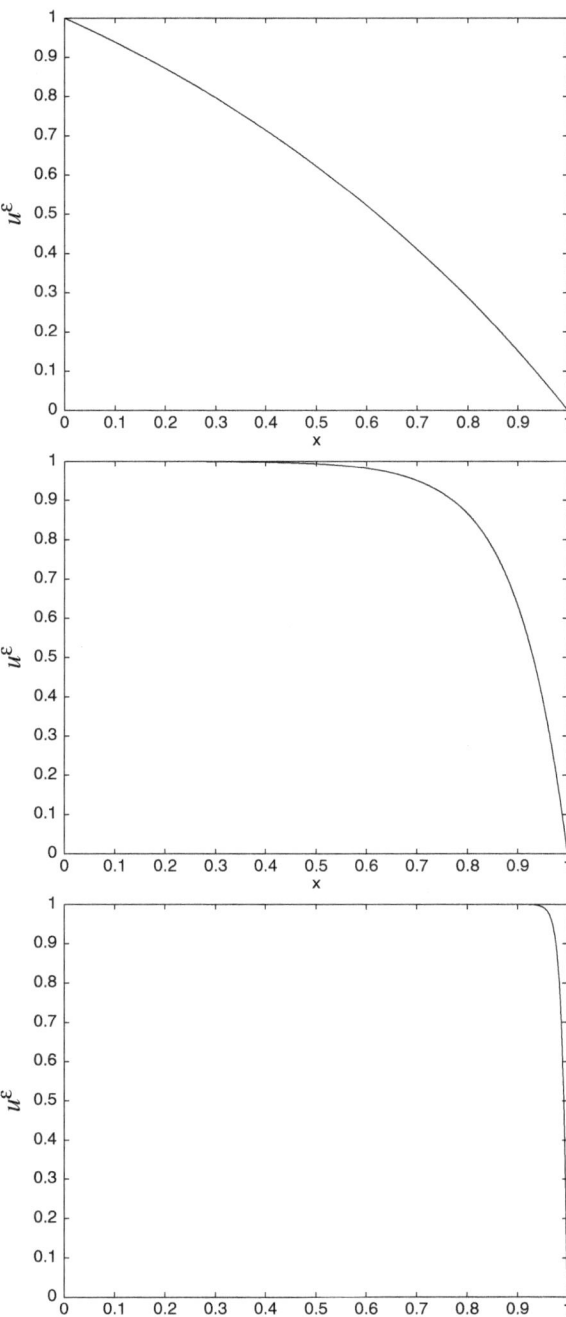

Remark 2.1 System (2.1) yields the simplest equations that capture essential features of fluid flow problems, where ε represents the fluid viscosity, the second derivative term models the diffusion, and the first derivative term models the advection. A reaction, "zeroth order term", could be added as well, see Sect. 2.3 and Chap. 4. In general, at the limit case, the equation change type, from elliptic to hyperbolic. The inflow boundary conditions are preserved, but the outflow conditions are lost.

Let us proceed with a straightforward Galerkin discretization of (2.1) using finite element method. We first rewrite (2.1) in a weak form, i.e, the exact solution

$$u^\epsilon \in V = \left\{ v \in H^1(0,1) \ : \ v(0) = 1 \text{ and } v(1) = 0 \right\},$$

satisfies

$$a(u^\epsilon, v) = \varepsilon \int_0^1 \frac{du^\epsilon}{dx} \frac{dv}{dx} \, dx + \int_0^1 \frac{du^\epsilon}{dx} v \, dx = 0 \quad \text{for all } v \in H_0^1(0,1). \tag{2.3}$$

Consider a discretization as described in (1.11), and let

$$V_h = \left\{ v_h \in V \ : \ v_h \text{ is linear in } (x_{j-1}, x_j) \text{ for } j = 1, \ldots, N+1 \right\},$$

$$V_0^h = \left\{ v_h \in H_0^1(0,1) \ : \ v_h \text{ is piecewise linear} \right\}.$$

The finite element approximation to u^ϵ is $u_h \in V_h$ such that

$$a(u_h, v) = 0 \quad \text{for all } v \in V_0^h. \tag{2.4}$$

Remark 2.2 Note that u_h depends on ε, although this is not explicitly indicated in the notation.

As depicted in Fig. 2.2, for a uniform mesh with $h = 1/16$ and $\varepsilon = 1$ the exact (2.2) and finite element (2.4) solutions seem quite close. However, for $\varepsilon = 0.01$ the finite element approximation presents spurious oscillations that are more prominent close to the boundary layer. Reducing the mesh to $h = 1/32$ improves the approximation, but the oscillations are still present. Refining the mesh even further would eventually result in an accurate approximation.

2.2.2 So, What Goes Wrong?

To better understand, or, at least, have a feeling of what goes wrong, we develop an error analysis for this problem.

We first investigate the continuity of the bilinear form $a(\cdot, \cdot)$. In fact, it follows from its definition and Cauchy–Schwartz inequality that

Fig. 2.2 For a uniform mesh with $h = 1/16$ we plot exact (*solid line*) and approximate (*dash-dot line*) solutions of (2.1) for $\varepsilon = 1$ (*top*) and $\varepsilon = 0.01$ (*center*). Note that the numerical solution is highly oscillatory close to the boundary layer at $x = 1$. For the same value of ε (= 0.01) a more refined mesh ($h = 1/32$) yields a numerical solution that is less oscillatory (*bottom*), but still unsatisfactory

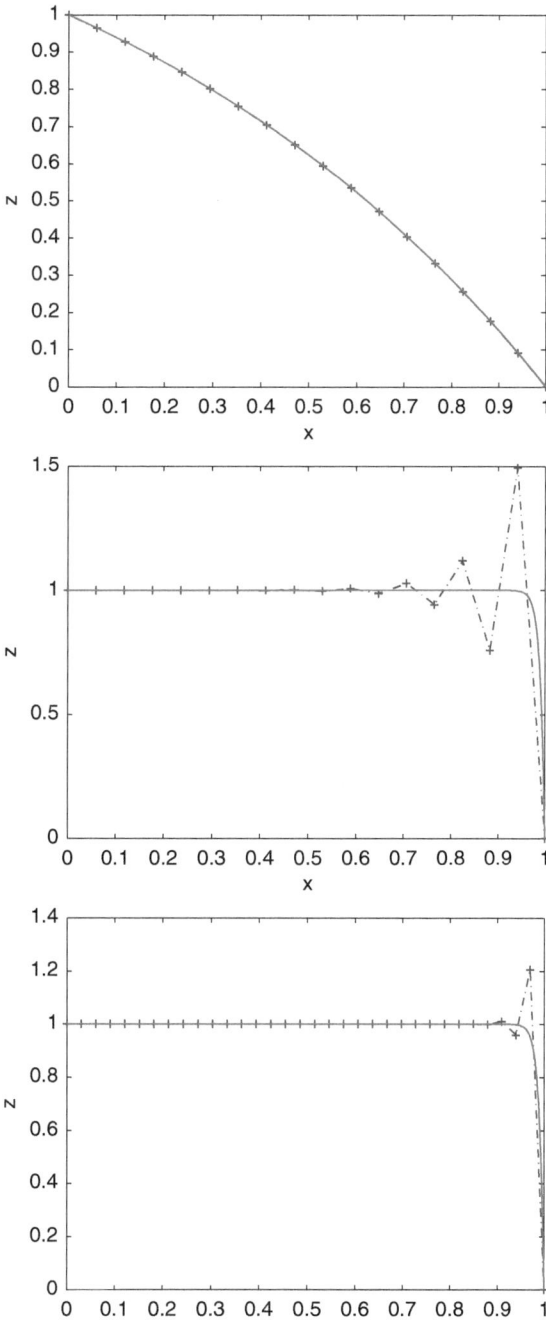

$$a(u, v) \leq c\|u\|_{H^1(0,1)}\|v\|_{H^1(0,1)} \quad \text{for all } u, v \in H_0^1(0, 1). \tag{2.5}$$

The difficulty starts when trying to derive the coercivity estimate:

$$a(v, v) = \varepsilon \int_0^1 \left(\frac{dv}{dx}\right)^2 dx + \int_0^1 \frac{dv}{dx} v \, dx$$

$$= \varepsilon \int_0^1 \left(\frac{dv}{dx}\right)^2 dx \geq C\varepsilon \|v\|_{H^1(0,1)}^2 \quad \text{for all } v \in H_0^1(0, 1), \tag{2.6}$$

since integration by parts yields $\int_0^1 (dv/dx)v \, dx = 0$, for $v \in H_0^1(0, 1)$. We also used Poincaré's inequality (Lemma 1.1) at the last step. So, estimate (1.16) yields

$$\|u^\varepsilon - u_h\|_{H^1(0,1)} \leq C\varepsilon^{-1}h|u^\varepsilon|_{H^2(0,1)}. \tag{2.7}$$

We stop now to try interpret the error estimate we just obtained. First of all, *there is convergence in h*. Indeed, for a fixed ε, the error goes to zero as the mesh size goes to zero. The problem is that the convergence in h is not uniform in ε. Hence, for ε small, unless the mesh size is also very small, the H^1 norm error estimate becomes large. The estimate is even worse than one can think at first glance, since $|u^\varepsilon|_{H^2(0,1)} = O(\varepsilon^{-3/2})$. This makes (2.7) and the traditional Galerkin method almost useless.

Another way to look at this problem is by first noticing that we would like to have

$$\lim_{\varepsilon \to 0} u_h = \lim_{\varepsilon \to 0} u^\varepsilon = 1.$$

After all, it would be just perfect to have a method that converges (with ε) to the correct solution *for a fixed mesh*. This is not happening. Indeed, looking at the matrix problem coming from (2.4), it is matter of computation to show that [146]

$$-\frac{\varepsilon}{h^2}(u_{j+1} - 2u_j + u_{j-1}) + \frac{1}{2h}(u_{j+1} - u_{j-1}) = 0, \quad u_0 = 1, \ u_{N+1} = 0, \tag{2.8}$$

where $u_j = u_h(x_j)$. Assume N even. At the $\varepsilon = 0$ limit, $u_{j+1} = u_{j-1}$. This and the boundary conditions originate the oscillatory behavior of the approximate solution. See Fig. 2.3.

Remark 2.3 Note that although we used a finite element scheme to derive (2.8), this scheme is also a finite difference scheme which uses a central difference approximation for the convective term du/dx. The more naive finite difference approximation

$$-\frac{\varepsilon}{h^2}(u_{j+1} - 2u_j + u_{j-1}) + \frac{1}{h}(u_j - u_{j-1}) = 0, \quad u_0 = 1, \ u_{N+1} = 0, \tag{2.9}$$

Fig. 2.3 Exact (*solid line*) and numerical (*dash-dot line*) solutions defined by (2.2) and (2.3) for $\varepsilon = 10^{-5}$ and $N = 16$. Note that the approximate solution has nodal values u_j close to zero for j odd and close to one for j even

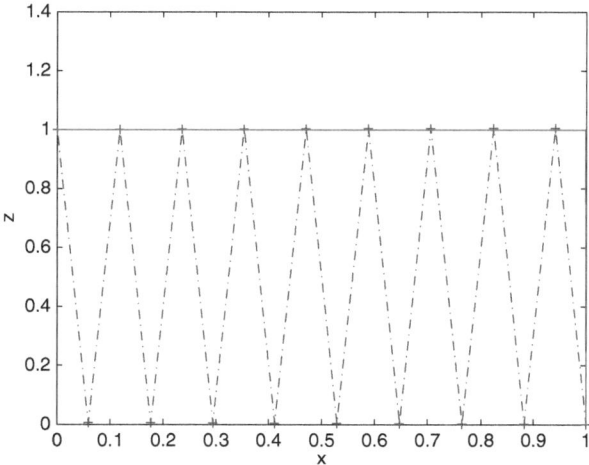

Fig. 2.4 Finite difference approximation (*dash-dot line*) with first order discretization given by (2.9) for the convection term, for $\varepsilon = 0.01$ and $h = 1/32$. Note the absence of oscillations

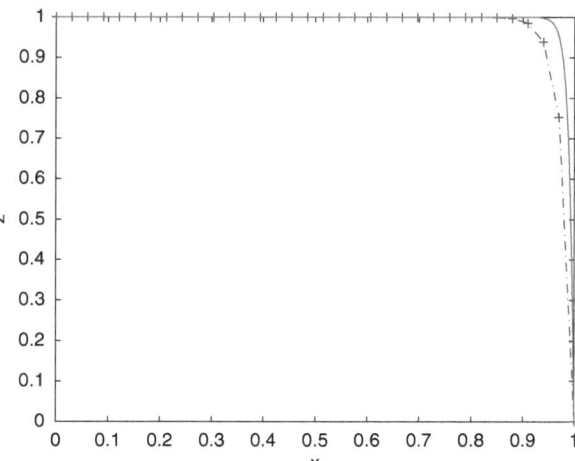

yields, however, a better result. See Fig. 2.4. In fact, for this scheme, $u_j = u_{j-1}$, as ε goes to zero. Since $u_0 = 1$, it holds that $u_j = 1$ in the $\varepsilon \to 0$ limit:

$$\lim_{\varepsilon \to 0} u_h(x_j) = \lim_{\varepsilon \to 0} u^\varepsilon(x_j) = 1, \quad \text{for } j = 1, \dots, N.$$

2.2.3 Matching Asymptotic Expansions

The behavior we described above is typical in singular perturbed PDEs, where the onset of boundary layers is a common phenomenon. But this is not all that can happen. For instance, in plate models, in particular for the Reissner–Mindlin equation, as the plate thickness goes to zero (that is, the small parameter in this case), numerical "locking" occurs, i.e., if a careless method is used, the computed solution goes to zero (a wrong limit) [93].

Several numerical methods try to somehow overcome these and other difficulties related to asymptotic limits. Some methods perform well for a certain asymptotic range, for instance by assuming $\varepsilon \ll 1$. Some other methods try to be performing for a broader range of parameters. See, for instance, [44, 101, 139, 186].

Looking at these difficulties (and their corresponding solutions!) it becomes more clear that it is important to have a full understanding of the solution's behavior. This is useful not only to help designing new numerical methods, but also to analyze and estimate old ones. A valuable analysis tool is the method of *matching asymptotics*, where the exact solution for a given singular perturbed PDE is expressed in terms of a *formal* power series with respect to a small parameter. We call it formal since we are not concerned with convergence at this point, and so we assume that all manipulations are valid from the mathematical point of view. We shall explain how an asymptotic expansion can be developed by looking at a simple example.

Consider problem (2.1), and the formal asymptotic series

$$u^\varepsilon \sim u^0 + \varepsilon u^1 + \varepsilon^2 u^2 + \cdots , \qquad (2.10)$$

and formally substitute it in (2.1). Then

$$\frac{du^0}{dx} + \varepsilon\left(-\frac{d^2 u^0}{dx^2} + \frac{du^1}{dx} \right) + \cdots + \varepsilon^i\left(-\frac{d^2 u^{i-1}}{dx^2} + \frac{du^i}{dx} \right) + \cdots = 0.$$

By comparing the different powers of ε, it is natural to require that

$$\frac{du^0}{dx} = 0, \qquad \frac{du^1}{dx} = \frac{d^2 u^0}{dx^2}, \qquad \ldots, \qquad \frac{du^i}{dx} = \frac{d^2 u^{i-1}}{dx^2}, \qquad \ldots. \qquad (2.11)$$

Then all functions u^i are constants.

From the boundary conditions in (2.1), it would be natural to impose $u^0(0) = 1$, $u^i(0) = 0$ for $i \geq 1$, and $u^i(1) = 0$ for all i. However, the equations in (2.11) are of first order, and only one boundary condition is to be imposed. We *choose* the inflow boundary condition at $x = 0$, and set then

$$u^0(0) = 1, \qquad u^i(0) = 0 \quad \text{for } i > 0.$$

Since u^i are constants, we conclude that $u^0 = 1$ and $u^i = 0$ for $i \geq 0$. Thus the formal series (2.10) simplifies to

$$u^\varepsilon \sim 1. \qquad (2.12)$$

Of course such expansion does not satisfy the boundary condition at $x = 1$, although it seems to deliver a good approximation at the interior of the domain. We correct the boundary discrepancy by introducing the *boundary corrector U*. We would like to have

$$-\varepsilon \frac{d^2 U}{dx^2} + \frac{dU}{dx} = 0, \quad U(0) = 0, \quad U(1) = -1.$$

Note that if we make the change of coordinates $\hat{\rho} = (1 - x)/\varepsilon$, and set $\hat{U}(\hat{\rho}) = U(1 - \varepsilon\hat{\rho})$, then

$$-\frac{d^2 \hat{U}}{d\hat{\rho}^2}(\hat{\rho}) - \frac{d\hat{U}}{d\hat{\rho}}(\hat{\rho}) = 0, \qquad \hat{U}(0) = 1.$$

Noting that we need another boundary condition for $\hat{U}$, we try to ensure a "local behavior" by imposing

$$\lim_{\hat{\rho} \to \infty} \hat{U}(\hat{\rho}) = 0.$$

Hence, $\hat{U}(\hat{\rho}) = -e^{-\hat{\rho}}$ and

$$U(x) = -e^{(x-1)/\varepsilon} \tag{2.13}$$

The asymptotic expansion then becomes

$$u^\varepsilon(x) \sim u^0 + \hat{U}(\hat{\rho}) = 1 - e^{(x-1)/\varepsilon}. \tag{2.14}$$

Although this is a very simple problem, some characteristics of asymptotic expansion are present here. First, the *outer expansion* (2.12) satisfies the operator, but fails to satisfy the boundary conditions. Then an *inner expansion* (2.13) has to be added. The terms of the inner expansion depend on ε through a change of coordinates, and become exponentially small in the interior of the domain. Finally, note that the expansion does not give the exact solution—indeed it fails to satisfy the boundary condition at $x = 0$ (an exponentially small miss). The idea is that, as $\varepsilon \to 0$, the expansion approximates the solution.

Remark 2.4 Here we could have multiplied U by a smooth cut-off function that values one in the vicinity of $x = 1$ (say, in $(1/2, 1)$), and zero at $x = 0$. Then the asymptotic expansion would satisfy all boundary conditions, at the expense of no longer satisfying the operator exactly. That would only introduce an exponential small error with respect to $1/\varepsilon$. For higher dimensional problems, the introduction of the cut-off function is "mandatory," as the boundary correctors are defined only in a neighborhood of the boundary, and not in the whole domain as here. See Chap. 4.

For the sake of comparison, the exact solution is

$$u^\epsilon(x) = 1 - \frac{e^{x/\epsilon} - 1}{e^{1/\epsilon} - 1} = 1 - \frac{e^{(x-1)/\epsilon} - e^{-1/\epsilon}}{1 - e^{-1/\epsilon}}$$

$$= 1 - (e^{(x-1)/\epsilon} - e^{-1/\epsilon})(1 + e^{-1/\epsilon} + e^{-2/\epsilon} + e^{-3/\epsilon} + \cdots)$$

$$= 1 - (e^{(x-1)/\epsilon} - e^{-1/\epsilon}) - s(e^{(x-1)/\epsilon} - e^{-1/\epsilon}) = 1 - e^{(x-1)/\epsilon} + r,$$

where

$$s = \frac{e^{-1/\epsilon}}{1 - e^{-1/\epsilon}}, \qquad r = e^{-1/\epsilon}\left(1 - \frac{e^{(x-1)/\epsilon} - e^{-1/\epsilon}}{1 - e^{-1/\epsilon}}\right).$$

Thus, the difference between the exact solution and the asymptotic expansion (2.14) is r, and, for all $x \in (0, 1)$, $|r| \le e^{-1/\epsilon}$ is exponentially small with respect to $1/\epsilon$.

2.3 A More General Singular Perturbed Second Order ODE

The asymptotic expansion of Sect. 2.2 is remarkably simple, but that does not represent the general case. Consider the differential operator

$$\mathscr{L}^\epsilon u^\epsilon = -\epsilon \frac{d^2 u^\epsilon}{dx^2} + \beta(x)\frac{du^\epsilon}{dx} + \sigma(x)u^\epsilon,$$

and the problem

$$\mathscr{L}^\epsilon u^\epsilon = f \quad \text{in } (0, 1), \tag{2.15}$$

$$u^\epsilon(0) = u^\epsilon(1) = 0. \tag{2.16}$$

We assume that $\epsilon > 0$, that β, σ, and f are smooth functions, and that both $\sigma + \beta/2 > 0$ and $\beta > 0$ are positive.

In Sect. 2.3.1 we shall develop asymptotic expansion for u^ϵ, and in Sect. 2.3.2 we show error estimates.

2.3.1 Asymptotic Expansion

Consider the asymptotic series

$$u^0 + \epsilon u^1 + \epsilon^2 u^2 + \cdots$$

and formally substitute it in (2.15). Then

$$\beta(x)\frac{du^0}{dx} + \sigma(x)u^0 + \varepsilon\left(-\frac{d^2u^0}{dx^2} + \beta(x)\frac{du^1}{dx} + \sigma(x)u^1\right) + \cdots$$

$$+ \varepsilon^i\left(-\frac{d^2u^{i-1}}{dx^2} + \beta(x)\frac{du^i}{dx} + \sigma(x)u^i\right) + \cdots = f.$$

Grouping the different powers of ε, we set

$$\mathscr{L}^0 u^0 = f, \qquad \mathscr{L}^0 u^1 = \frac{d^2u^0}{dx^2}, \qquad \ldots, \qquad \mathscr{L}^0 u^i = \frac{d^2u^{i-1}}{dx^2}, \qquad \ldots,$$

(2.17)

where $\mathscr{L}^0 v = \beta(x)dv/dx + \sigma(x)v$.

Since the equations in (2.17) are of first order, it is not possible to impose $u^i(0) = u^i(1) = 0$, conforming (2.16). We then set

$$u^i(0) = 0.$$

We correct the discrepancy at $x = 1$ by introducing the boundary corrector U. We would like to have

$$\mathscr{L}^\varepsilon U = 0, \quad U(0) = 0, \quad U(1) = u^0(1) + \varepsilon u^1(1) + \varepsilon^2 u^2(1) + \cdots$$

From the change of coordinates $\hat{\rho} = \varepsilon^{-1}(1-x)$, and making $\hat{U}(\hat{\rho}) = U(1 - \varepsilon\hat{\rho})$, it follows that

$$-\frac{d^2\hat{U}}{d\hat{\rho}^2}(\hat{\rho}) - \beta(1-\varepsilon\hat{\rho})\frac{d\hat{U}}{d\hat{\rho}}(\hat{\rho}) + \varepsilon\sigma(1-\varepsilon\hat{\rho})\hat{U}(\hat{\rho}) = 0,$$

$$\hat{U}(0) = u^0(1) + \varepsilon u^1(1) + \varepsilon^2 u^2(1) + \cdots$$

Going one step further, we develop the Taylor expansions

$$\beta(1-\varepsilon\hat{\rho}) = \beta(1) - \varepsilon\hat{\rho}\frac{db}{dx}(1) + \frac{\varepsilon^2\hat{\rho}^2}{2}\frac{d^2b}{dx^2}(1) - \cdots,$$

$$\sigma(1-\varepsilon\hat{\rho}) = \sigma(1) - \varepsilon\hat{\rho}\frac{dc}{dx}(1) + \frac{\varepsilon^2\hat{\rho}^2}{2}\frac{d^2c}{dx^2}(1) - \cdots.$$

Finally, assuming the asymptotic expansion

$$\hat{U} \sim \hat{U}^0 + \varepsilon\hat{U}^1 + \varepsilon^2\hat{U}^2 + \cdots,$$

we gather that

$$-\frac{d^2\hat{U}^0}{d\hat{\rho}^2} - \beta(1)\frac{d\hat{U}^0}{d\hat{\rho}} = 0,$$

$$U^0(0) = u^0(1).$$

Noting that we need another boundary condition for $\hat{U}^0$, we try to ensure a "local behavior" by imposing

$$\lim_{\hat{\rho}\to\infty} \hat{U}^0(\hat{\rho}) = 0.$$

Hence, $\hat{U}^0(\hat{\rho}) = u(1) \exp\left(-\beta(1)\hat{\rho}\right)$.

Similarly,

$$-\frac{d^2\hat{U}^1}{d\hat{\rho}^2} - \beta(1)\frac{d\hat{U}^1}{d\hat{\rho}} = -\hat{\rho}\frac{d\beta}{dx}(1)\frac{d\hat{U}^0}{d\hat{\rho}} - \sigma(1)\hat{U}^0,$$

$$\hat{U}^1(0) = u^1(1), \qquad \lim_{\hat{\rho}\to\infty} \hat{U}^1(\hat{\rho}) = 0,$$

etc. It is possible to show that for every positive integer i there exist ε-independent positive constants c and α such that

$$\frac{d\hat{U}^i}{d\hat{\rho}}(\hat{\rho}) + \hat{U}^i(\hat{\rho}) \le c\exp(-\alpha\hat{\rho}). \tag{2.18}$$

So, putting everything together, we have that

$$u^\varepsilon(x) \sim u^0(x) + \varepsilon u^1(x) + \varepsilon^2 u^2(x) + \cdots$$
$$- \hat{U}^0(\varepsilon^{-1}(1-x)) - \varepsilon\hat{U}^1(\varepsilon^{-1}(1-x)) - \varepsilon^2\hat{U}^2(\varepsilon^{-1}(1-x)) - \cdots . \tag{2.19}$$

By construction, the above infinite power series *formally* solves the ODE (2.15). We did not make any comment regarding convergence of the above expansion. Actually, what we will prove is that a *truncated expansion* approximates well the exact solution.

Remark 2.5 Note that each term u^i in the series (2.19) is independent of ε, and each boundary corrector terms U^i depends on ε but only up to a change of coordinates. Also, U^i does not satisfy the boundary condition at $x = 0$, but this error is exponentially small. As pointed out in Remark 2.4, it is possible to correct this by introducing a smooth cut-off function; see also Remark 2.9.

2.3.2 Truncation Error Analysis

We start by developing here an analysis quite similar to that of Sect. 2.2.2. To simplify the computations, we assume here that the functions β and $\sigma \ge 0$ are actually positive constants—otherwise it would be necessary to take into account the effect of replacing them by their truncated Taylor expansions.

We first obtain stability estimates.

Lemma 2.1 *If $\mathscr{L}^\epsilon v = g$ weakly, and $v \in H_0^1(0, 1)$, then*

$$\|v\|_{H^1(0,1)} \leq c\varepsilon^{-1}\|g\|_{H^{-1}(0,1)}.$$

Proof From (2.6), we conclude that

$$\|v\|_{H^1(0,1)}^2 \leq c\varepsilon^{-1}a(v, v) = c\varepsilon^{-1}(g, v) \leq c\varepsilon^{-1}\|g\|_{H^{-1}(0,1)}\|v\|_{H^1(0,1)}.$$

$\square$

Corollary 2.1 *If $w \in H^1(0, 1)$ is the weak solution of*

$$\mathscr{L}^\epsilon w = g, \qquad w(0) = w_0, \quad w(1) = w_1,$$

then

$$\|w\|_{H^1(0,1)} \leq c\varepsilon^{-1}\left(\|g\|_{H^{-1}(0,1)} + |w_0| + |w_1|\right).$$

Proof Take any function $w_{bc} \in H^1(0, 1)$ be such that $w_{bc}(0) = w_0$, and $w_{bc} = w_1$, with $\|w_{bc}\|_{H^1(0,1)} \leq c(|w_0| + |w_1|)$. Then Lemma 2.1 with $v = w - w_{bc}$ yields the result since $\|w\|_{H^1(0,1)} \leq \|w - w_{bc}\|_{H^1(0,1)} + \|w_{bc}\|_{H^1(0,1)}$. $\square$

Lemma 2.2 *Let $\mathscr{L}^0 v = g$ in $(0, 1)$. Then there exists a constant c depending on β, σ, and g such that*

$$\|\frac{d^2 v}{dx^2}\|_{L^\infty(0,1)} + \|\frac{dv}{dx}\|_{L^\infty(0,1)} + \|v\|_{L^\infty(0,1)} \leq c.$$

Proof Note that since β is always positive, v solves $dv/dx + pv = \bar{g}$ with $p = \sigma/\beta$ and $\bar{g} = g/\beta$. Then [33]

$$u(x) = \frac{\int_0^x \mu(s)\bar{g}(s)\,ds}{\mu(x)}, \qquad \mu(x) = \exp\int_0^x p(s)\,ds.$$

Thus, $\|\mu\|_{L^\infty(0,1)} \leq \exp\|p\|_{L^\infty(0,1)}$ and $|\mu(x)| \geq \exp(-\|p\|_{L^\infty(0,1)})$ for all $x \in (0, 1)$. So, $\|v\|_{L^\infty(0,1)} \leq \exp(2\|p\|_{L^\infty(0,1)})\|\bar{g}\|_{L^\infty(0,1)}$. The bound for dv/dx follows from $\|dv/dx\|_{L^\infty(0,1)} = \|\bar{g} - pv\|_{L^\infty(0,1)}$. The bound for $\|d^2v/dx^2\|_{L^\infty(0,1)}$ follows from the previous arguments, since $w = dv/dx$ solves $dw/dx + pw = \bar{g} - (dp/dx)v$. $\square$

From (2.17) and the above lemma, we find that for all $i \in \mathbb{N}$, there exists a constant c (depending on i, f, β, and σ) such that

$$\left\|\frac{d^2 u_i}{dx^2}\right\|_{L^\infty(0,1)} + \left\|\frac{du_i}{dx}\right\|_{L^\infty(0,1)} + \|u_i\|_{L^\infty(0,1)} \leq c. \tag{2.20}$$

With the aid of Corollary 2.1, we are ready to estimate how well the asymptotic expansion approximates the exact solution of (2.15), (2.16). Let

$$e_N(x) = u^\epsilon(x) - \sum_{i=0}^{N} \varepsilon^i u^i(x) + \sum_{i=0}^{N} \varepsilon^i \hat{U}^i(\varepsilon^{-1}(1-x)),$$ (2.21)

From its construction, $e_N \in H^1(0,1)$, and

$$\mathcal{L}^\epsilon e_N = \varepsilon^{N+1}\frac{du^i}{dx^2}, \qquad e_N(0) = \sum_{i=0}^{N} \varepsilon^i \hat{U}^i(\varepsilon^{-1}), \quad e_N(1) = 0.$$ (2.22)

Using now Corollary 2.1, Eqs. (2.20), (2.22), and (2.18), we gather that there exists a constant c such that

$$\|e_N\|_{H^1(0,1)} \le c\varepsilon^N.$$

This estimate is not sharp. We improve it by adding and subtracting the $(N+1)$th term of the expansion:

$$\|e_N\|_{H^1(0,1)} \le \|e_{N+1}\|_{H^1(0,1)} + \|e_N - e_{N+1}\|_{H^1(0,1)}$$
$$\le c\big[\varepsilon^{N+1} + \varepsilon^{N+1}\|u^{N+1}\|_{H^1(0,1)} + \varepsilon^{N+1}\|\hat{U}^{N+1}(\varepsilon^{-1}(1-\cdot))\|_{H^1(0,1)}\big]$$
$$\le c\varepsilon^{N+1/2}.$$ (2.23)

The last estimate was possible since

$$\|\hat{U}^{N+1}(\varepsilon^{-1}(1-\cdot))\|_{H^1(0,1)}^2$$
$$= \int_0^1 \left|\frac{d\hat{U}^{N+1}}{dx}(\varepsilon^{-1}(1-x))\right|^2 + \left|\hat{U}^{N+1}(\varepsilon^{-1}(1-x))\right|^2 dx$$
$$\le \varepsilon \int_0^\infty \varepsilon^{-2}\left|\frac{d\hat{U}^{N+1}}{dx}(\hat{\rho})\right|^2 + \left|\hat{U}^{N+1}(\hat{\rho})\right|^2 d\hat{\rho}$$
$$\le c\varepsilon \int_0^\infty \varepsilon^{-2}\exp(-2\alpha\hat{\rho})\,d\hat{\rho} \le c\varepsilon^{-1}.$$

Estimates in other norms can be obtained in a similar fashion:

$$\|e_N\|_{L^2(0,1)} \le \|e_{N+1}\|_{H^1(0,1)} + \|e_N - e_{N+1}\|_{L^2(0,1)} \le c\varepsilon^{N+1}.$$

Remark 2.6 It is also possible to derive interior error estimates, by considering the interval $I_\delta = (0, 1-\delta)$, for $\delta \in (0, 1/2)$. Then compute $\|e_N\|_{H^1(I_\delta)}$ and $\|e_N\|_{L^2(I_\delta)}$, proceeding as in (2.23). We leave the details to the reader.

Interior estimates are important because global estimates in H^1 norms are dominated by boundary layer terms (high derivatives). Such terms, however, are exponentially small in the interior of the domain, allowing for better convergence rates away from the boundary. No improvements are obtained in L^2 norms.

We obtained then the following important result.

Theorem 2.1 *Let u^ϵ be the solution of the ODE (2.15), and let e_N be as in (2.21). Then, for every non-negative integer N, there exists a constant c such that*

$$\|e_N\|_{H^1(0,1)} \le c\varepsilon^{N+1/2}. \qquad \|e_N\|_{L^2(0,1)} \le c\varepsilon^{N+1}, \qquad \|e_N\|_{H^1(I_\delta)} \le c_\delta\varepsilon^{N+1},$$

where I_δ is defined in Remark 2.6. The constant c might depend on N, f, β, and σ, but not on ε. The constant c_δ might also depend on δ.

Remark 2.7 Theorem 2.1 does not imply convergence of the power series as N goes to infinity, since the constants that appear in the right-hand side of the estimates depend on N. What the theorem provides is a convergence in ε, i.e., if ε is quite small, then the asymptotic expansion truncation error gets small as well.

Remark 2.8 The asymptotic *rule of the thumb* works: the error estimates present in Theorem 2.1 are of the same order as the terms left out of the truncated asymptotic expansion.

Remark 2.9 It is possible to derive an error estimate under the presence of a cut-off function χ using the following result.

Lemma 2.3 *For every non-negative integer i, there exist ε-independent positive constants C and α such that*

$$\sup_{x \in (0,1)} \mathscr{L}^\epsilon \left(\chi(x) U^i(\varepsilon^{-1}(1-x)) \right) \le C \exp(\alpha\varepsilon^{-1})$$

Proof It follows from the definition of U^i that

$$\mathscr{L}^\epsilon \left(\chi U^i \right) = \mathscr{L}^\epsilon \left(\chi U^i \right) - \chi \mathscr{L}^\epsilon \left(U^i \right) = -\varepsilon \left(\chi U^i \right)'' + \beta(x) \left(\chi' U^i \right) + \varepsilon \chi \left(U^i \right)''$$

$$= -\varepsilon \chi'' U^i - 2\varepsilon \chi' U^{i'} + \beta(x) \left(\chi' U^i \right).$$

The result follows from the definition of χ and estimate (2.18). $\qquad\square$

2.4 Conclusions

In general, it is not easy to design numerical methods schemes for singular perturbed PDEs. The layers that characterize such problems cause numerical instabilities that are hard to tame. Understanding such layers is then important, and there is no easier way to do so than considering one-dimensional cases.

Asymptotic expansions, even when they are formal, allow a clear description of the solution. It is also important to note that the derivation of the expansion might be formal, but its outcome can be justified rigorously, by estimating the difference between the exact solution and truncated expansions. Such estimates do not guarantee that the series converge. Instead, it tells us that the first few terms of the expansion yield a remarkable approximation to the exact solution. And that is enough for most applications.

Chapter 3
An Application in Neuroscience: Heterogeneous Cable Equation

Abstract We consider here a simplified version of an equation that models the voltage transmission along neurons, modeled here by a "cable." If the thickness of the cable is much smaller than its length, it originates a singular perturbed behavior equation that does not differ substantially from what was investigated in Chap. 2. Moreover, other interesting asymptotics arise when considering a large number of synapses.

We also show here that the Multiscale Finite Element Method yields good approximations under all asymptotic regimes, even when the Galerkin Method fails.

3.1 Introduction

Most problems in neuroscience are of multiscale type by nature, due to the incredibly complex interplay of several biochemical elements that influence the behavior of cells and the whole brain. Of current particular interest is the modeling of large networks of neurons, incorporating physiological details of each neuron. It is then only natural that multiscale techniques were explored from different points of view, such as modeling [66, 68, 127, 166], numerical and computational aspects [27, 32, 48, 120, 129–132, 179, 182, 183, 191, 200], and asymptotics [36, 37, 46, 91, 153, 163, 165, 173, 187, 190].

The problem we consider involves modeling dendrites with an arbitrary distribution of synapses. Such problem can be homogenized [165] with techniques not different from what we consider here. Such process "replaces" the heterogeneous dendrites with a homogeneous one that has similar physical behavior. One has to keep in mind, however, that homogenization is valid only under certain assumptions, as periodicity. Also, the number of synapses has to be large. Such assumptions are often questionable [32, 165].

With such motivation, we also investigate the cable equation numerically, proposing a multiscale finite element method that is asymptotically robust. Our results are based on [158].

The transient model that we are based on is derived in [193, 194], but here we consider its steady-state version, posed in an interval. After a change of coordinate,

© The Author(s) 2017
A.L. Madureira, *Numerical Methods and Analysis of Multiscale Problems*,
SpringerBriefs in Mathematics, DOI 10.1007/978-3-319-50866-5_3

the final mathematical problem is to find a function V such that

$$-\epsilon \frac{\partial^2 V}{\partial x^2} + V + GV = f \quad \text{in } (0, 1),$$
$$V(0) = V(1) = 0.$$

$$\text{(3.1)}$$

In the equation above, ϵ is a positive constant, and

$$G = \sigma^{\text{in}} + \sigma^{\text{ex}}, \qquad f = \sigma^{\text{in}} V^{\text{in}} + \sigma^{\text{ex}} V^{\text{ex}},$$

where the constants V^{ex} and V^{in} are the reversal potentials. The inhibitory and excitatory synapses are modeled by

$$\sigma^{\text{in}} = \sum_{l=1}^{N^i} g_l^i \delta_{x_l^i}, \qquad \sigma^{\text{ex}} = \sum_{l=1}^{N^e} g_l^e \delta_{x_l^e}, \tag{3.2}$$

where $\delta_{x_l^i}$ are Dirac deltas located at $x_l^i \in (0, 1)$ with constant strengths g_l^i, for $l = 1, \ldots, N^i$. Similar notation holds for the excitatory synapses. The Dirac delta δ_{x^*} is defined by its action over any continuous function g by Haroske and Triebel [125]

$$\int_0^1 \delta_{x^*} g(x)\, dx = g(x^*). \tag{3.3}$$

   Note that (3.1) is a reaction-diffusion type equation, similar to (1.3) with $b = 0$, and with the added Dirac deltas, which cause the onset of internal boundary layers, making the problem even harder. Some asymptotic regimes that might cause numerical troubles include big differences in the strengths of the synapses, large number of synapses at arbitrary locations, and $\varepsilon \ll 1$. In such cases, the computational costs might soar, in particular when considering a large tree of dendrites with several branches (each branch is a line segment).
   Classical methods do not work well under such different regimes, and modern variants of finite element methods come as a viable option of discretization [13, 73, 83, 101, 102, 128]. We explore here the Multiscale Finite Element Method (MsFEM) [82, 83, 87, 88, 134, 136, 156], discussed in Sect. 1.3.2. The basic idea behind the method is to solve first local problems and extract microscale information. Such information is then upscaled into a homogenized global problem. The method that is not only accurate, but also computationally competitive in terms of costs [136]. Also, for steady problems in one-dimensional domains, the MsFEM is *nodally exact solution*; see Sect. 3.3.
   In the next two sections, we present the classical and multiscale finite element method applied to (3.1). We then investigate the model under different asymptotic limits in Sect. 3.4, and conclude with some final remarks in Sect. 3.5.

3.2 Classical Finite Element Method

The variational formulation associated with (3.1) is that $V \in H_0^1(0,1)$ satisfies

$$\int_0^1 \left(\epsilon \frac{\partial V}{\partial x} \frac{\partial w}{\partial x} + Vw + GVw \right) dx = \int_0^1 fw \, dx \quad \text{for all } w \in H_0^1(0,1). \tag{3.4}$$

Using (3.2), and the definition of Dirac deltas (3.3), it follows that

$$\int_0^1 GVw \, dx = \sum_{l=1}^{N^i} g_l^i V(x_l^i) w(x_l^i) + \sum_{l=1}^{N^e} g_l^e V(x_l^e) w(x_l^e),$$

$$\int_0^1 fw \, dx = V^{\text{in}} \sum_{l=1}^{N^i} g_l^i w(x_l^i) + V^{\text{ex}} \sum_{l=1}^{N^e} g_l^e w(x_l^e). \tag{3.5}$$

As in Sect. 1.2, we consider the nodal points

$$0 = x_0 < x_1 < \cdots < x_{N+1} = 1, \tag{3.6}$$

and the corresponding partition of $(0,1)$ into elements (x_{k-1}, x_k). Assume also that V can be approximated by continuous piecewise linear functions,

$$V(x) \approx V_h(x) = \sum_{k=1}^{N} V_k \psi_k(x), \tag{3.7}$$

where $V_1, \ldots, V_N$ are the unknowns, and ψ_k is a continuous piecewise linear basis function satisfying $\psi_k(x_j) = \delta_{k,j}$; see a typical basis function in Figs. 1.2 and 3.1.

To compute the unknowns, we use (3.4) replacing V by V_h and w by ψ_j, obtaining the N-dimensional system of equations:

$$\sum_{k=1}^{N} V_k \int_0^1 \left[\epsilon \frac{\partial \psi_k(x)}{\partial x} \frac{\partial \psi_j(x)}{\partial x} + \psi_k(x)\psi_j(x) + G\psi_k(x)\psi_j(x) \right] dx = \int_0^1 f\psi_j(x) \, dx, \tag{3.8}$$

for $j = 1, \ldots, N$.

Note that finite element formulations deal with evaluations of delta functions naturally, since it is based on variational forms, i.e., forms that involve computations of integrals (3.4). It is then easy to compute the action of the Dirac deltas, as in (3.5).

However, as we show latter, see, for instance, Sect. 3.4, the classical method does not yield accurate results independently of the parameters. In particular if $\varepsilon \ll 1$, we observe spurious oscillation, as in Fig. 3.2.

Fig. 3.1 Multiscale basis functions shown in *continuous line* for large (*top figure*) and small (*bottom figure*) ε. There are three inhibitory (location marked with ×), and two excitatory (location marked with ○) synapses. The classical basis functions are in dashes, and are shown to highlight the different behavior of the two approaches (classical versus multiscale)

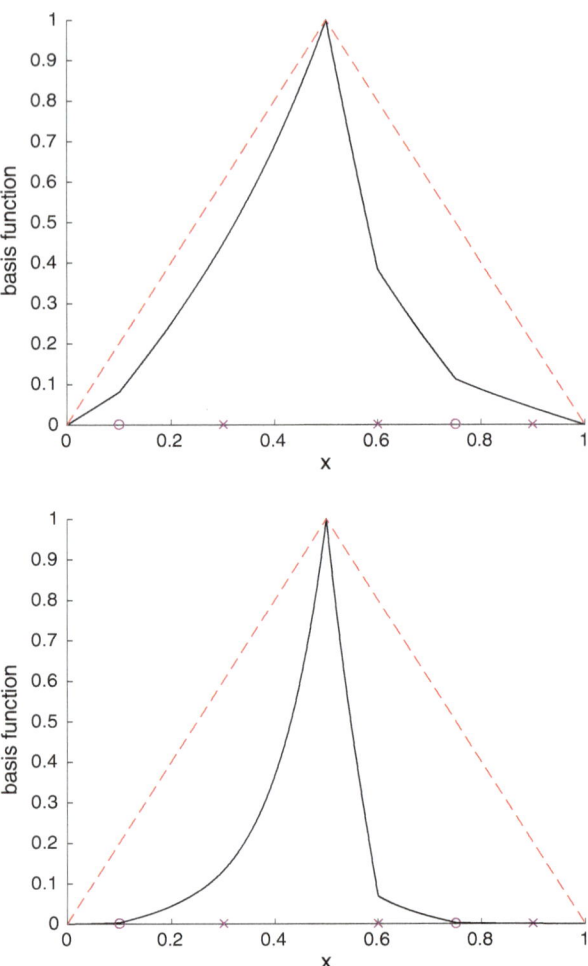

3.3 The Multiscale Finite Element Method

We employ here the Multiscale Finite Element Method (MsFEM), as described in Sect. 1.3.2, recalling that in the MsFEM, the approach is still based on finite elements, but with modified basis functions.

Considering the partition (3.6), we approximate V by the multiscale function

$$V_h^{\mathrm{ms}}(x) = \sum_{k=1}^{N} V_k^{\mathrm{ms}} \lambda_k(x). \tag{3.9}$$

Fig. 3.2 Exact and numerical solutions for small ϵ with values 2.5×10^{-4} (*top*) and 5×10^{-5} (*bottom*). The locations of the deltas are indicated with $\times$ (inhibitory) and $\circ$ (excitatory). The exact solutions are plotted using *continuous line*. The values of the classical piecewise linear Galerkin solutions are indicated by $*$, and the values of the multiscale solutions are indicated by $\square$

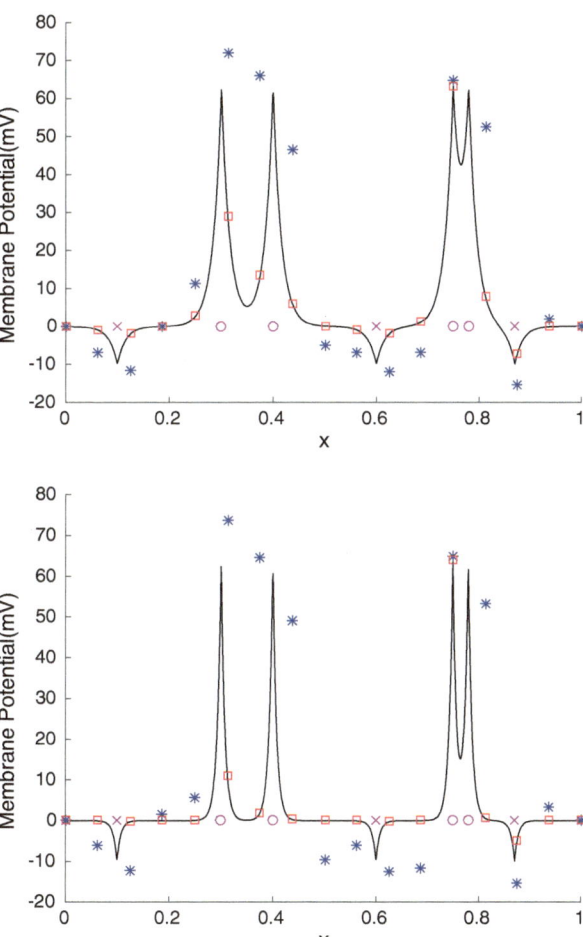

The unknowns of the above summation are the scalars $V_1^{\mathrm{ms}}, \ldots, V_N^{\mathrm{ms}}$. The multiscale basis functions $\lambda_1, \ldots, \lambda_N$ are still continuous, but not necessarily linear within each element, and satisfy the local problems

$$\lambda_k(x) = 0 \quad \text{if } x \notin (x_{k-1}, x_{k+1}),$$

$$-\epsilon \frac{\partial^2 \lambda_k}{\partial x^2} + \lambda_k + G\lambda_k = 0 \quad \text{in } (x_{k-1}, x_k) \text{ and } (x_k, x_{k+1}), \qquad (3.10)$$

$$\lambda_k(x_k) = 1,$$

for $k \in \{1, \ldots, N\}$.

In Fig. 3.1 we consider basis function for two different regimes, depending on if ε is large or small. The element is scaled to size one, and in both examples there are three inhibitory (location marked with $\times$) and two excitatory (marked with $\circ$)

synapses. Note that the multiscale basis function (continuous line) adapts not only to the location of the Dirac deltas, but also to the value of ε. Wherever there is a delta, the derivative of the basis function becomes discontinuous, and for small ε, the profile of basis function is exponential by parts. This is exactly the behavior of the solutions for (3.1), which the basis functions capture. Of course, the classical piecewise linear basis functions (dashed line) are oblivious to change of parameters.

The computation (3.10) of the multiscale basis functions λ_k is a *pre-processing step*. After that, the unknowns $V_1^{ms}, \ldots, V_N^{ms}$ are defined by (compare with the classical scheme (3.8)):

$$\sum_{k=1}^{N} V_k^{ms} \int_0^1 \left[\epsilon \frac{\partial \lambda_k(x)}{\partial x} \frac{\partial \lambda_j(x)}{\partial x} + \lambda_k(x)\lambda_j(x) + G\lambda_k(x)\lambda_j(x) \right] dx = \int_0^1 f\lambda_j(x)\, dx,$$

$$(3.11)$$

for $j = 1, \ldots, N$. See [158] for details on how to compute the basis functions for this particular problem, and a discussion regarding the efficiency of the method.

As shown in Sect. 1.3.2, the multiscale method yields exact solution for linear one-dimensional elliptic problems. We repeat here the proof for the present case. We need to show that the interpolation of V given by $I_h V = \sum_{k=1}^{N} V(x_k)\lambda_k$ solves

$$\int_0^1 \left[\epsilon \frac{\partial I_h V}{\partial x} \frac{\partial \lambda_j}{\partial x} + I_h V \lambda_j + G I_h V \lambda_j \right] dx = \int_0^1 f\lambda_j\, dx$$

$$= \int_0^1 \left[\epsilon \frac{\partial V}{\partial x} \frac{\partial \lambda_j}{\partial x} + V\lambda_j + GV\lambda_j \right] dx \quad \text{for } j = 1, \ldots, N. \qquad (3.12)$$

From an integration by parts, (3.10), and the identity $I_h V = V$ at every node, it follows that

$$\int_{x_{i-1}}^{x_i} \left[\epsilon \frac{\partial (I_h V - V)}{\partial x} \frac{\partial \lambda_j}{\partial x} + (I_h V - V)\lambda_j + G(I_h V - V)\lambda_j \right] dx = 0,$$

at every element (x_{i-1}, x_i). Thus, (3.12) holds.

3.4   Robustness of the Multiscale Method

In what follows, we consider two different asymptotic limits, that is, when ε gets small, and the case of a large number of deltas. We consider the behavior of the exact solution, and compare how the classical and multiscale finite element methods perform under these extreme situations.

3.4.1   Low Diffusion Regime

Consider $\epsilon \ll 1$. Then, using matching asymptotics, we postulate that

$$V(x) \sim V_0(x) + \epsilon V_1(x) + \epsilon^2 V_2(x) + \cdots, \qquad (3.13)$$

where the functions V_j are to be determined. Formally replacing (3.13) in (3.1), we gather that

$$V_0+GV_0+\epsilon\left(-\frac{\partial^2 V_0}{\partial x^2}+V_1+GV_1\right)+\epsilon^2\left(-\frac{\partial^2 V_1}{\partial x^2}+V_2+GV_2\right)+\cdots=f \quad \text{in } (0,1).$$

Collecting the $\epsilon=0$ limit terms, we see that $V_0+GV_0=f$, that is,

$$V_0+\left(\sum_{l=1}^{N^i} g_l^i \delta_{x_l^i}+\sum_{l=1}^{N^e} g_l^e \delta_{x_l^e}\right)V_0=\sum_{l=1}^{N^i} g_l^i \delta_{x_l^i}V^{\text{in}}+\sum_{l=1}^{N^e} g_l^e \delta_{x_l^e}V^{\text{ex}}.$$

Multiplying by a continuous function ϕ and integrating in $(0,1)$, we gather that

$$\int_0^1 V_0\phi\,dx + \sum_{l=1}^{N^i} g_l^i V_0(x_l^i)\phi(x_l^i)+\sum_{l=1}^{N^e} g_l^e V_0(x_l^e)\phi(x_l^e)$$

$$= V^{\text{in}}\sum_{l=1}^{N^i} g_l^i \phi(x_l^i) + V^{\text{ex}}\sum_{l=1}^{N^e} g_l^e \phi(x_l^e).$$

Then

$$V_0(x)=\begin{cases} 0 & \text{if } x \notin \{x_1^i,\ldots,x_{N^i}^i, x_1^e,\ldots,x_{N^e}^e\}, \\ V^{\text{in}} & \text{if } x \in \{x_1^i,\ldots,x_{N^i}^i\}, \\ V^{\text{ex}} & \text{if } x \in \{x_1^e,\ldots,x_{N^e}^e\}. \end{cases}$$

Hence, we formally show that V approaches V_0 as $\epsilon \to 0$. Since V_0 is discontinuous and V is continuous, there is an onset of internal layers at the points of discontinuity (the delta locations), a source of numerical difficulties [186].

Consider the numerical tests presented in Fig. 3.2. In both cases, there are three inhibitory (locations marked with ×) and three excitatory (∘) synapses, $V^{\text{in}} = -10$, $V^{\text{ex}} = 65$, $g_l^i = 8 \times 10^{-2}$, $g_l^e = 10^{-2}$. Also, $\epsilon = 2.5 \times 10^{-4}$ (top figure) and $\epsilon = 5 \times 10^{-5}$ (bottom figure). The "exact solutions" (computed with a very refined mesh) is displayed in solid line. We also computed the classical Galerkin solutions and the multiscale solutions with the same uniform mesh, where each element has length 2^{-4}. The nodal values of the classical Galerkin solution are plotted with ∗, and the nodal values of the multiscale solution are plotted with □.

Note that the exact solution is close to zero away from the deltas. This effect becomes more drastic as $\varepsilon \to 0$. Over the deltas the exact solution is close to either $V^{\text{in}} = -10$ or $V^{\text{ex}} = 65$. This confirms numerically that $V \to V_0$ as $\varepsilon \to 0$, as predicted by the theory. In terms of the numerical solutions, the classical method misses completely the exact solution, while the multiscale solution is nodally exact, as it should be.

3.4.2 Large Number of Synapses

We investigate now what happens as the number of deltas becomes large. For simplicity, suppose that $N^i = N^e$, and define $\alpha = 1/(2N^i)$. Assume further that the synapses are disposed periodically, i.e., the Dirac deltas are located at the sites $x_l^i = (2l - 1)\alpha$ and $x_l^e = 2l\alpha$. In the present case, the interest is when the synapses are narrowly packed, i.e. $\alpha \ll 1$, and this situation is tricky to analyze.

To find out how the solution V depends on the parameter α, we use arguments as in [21, Appendix A]; see also [51].

The idea is to characterize V as the minimizer of the energy functional

$$J(V) + \alpha^{-1} I_\alpha(V), \tag{3.14}$$

where

$$J(V) = \frac{1}{2} \int_0^1 \epsilon \left(\frac{\partial V}{\partial x} \right)^2 + V^2 \, dx,$$

$$I_\alpha(V) = \frac{\alpha}{2} \sum_{l=1}^{N^i} g_l^i V^2(x_l^i) + \frac{\alpha}{2} \sum_{l=1}^{N^e} g_l^e V^2(x_l^e)$$

$$- \alpha V^{in} \sum_{l=1}^{N^i} g_l^i V(x_l^i) - \alpha V^{ex} \sum_{l=1}^{N^e} g_l^e V(x_l^e).$$

Assume that $g_l^i = g^i(x_l^i)$ and $g_l^e = g^e(x_l^e)$ for some functions g^i and g^e defined in $(0, 1)$, having at most a finite number of discontinuities. Then, as $\alpha \to 0$, the term I_α concentrates most of the total energy. Thus $\lim_{\alpha \to 0} V = W_0$ in $L^2(0, 1)$, where W_0 minimizes $\lim_{\alpha \to 0} I_\alpha$. Note that, for a given function W, $\lim_{\alpha \to 0} I_\alpha(W) = I_0(W)$, where

$$I_0(W) = \frac{1}{2} \int_0^1 (g^i + g^e) V^2 \, dx - \int_0^1 (V^{in} g_l^i + V^{ex} g^e) V \, dx.$$

Since W_0 is the argument that minimizes $I_0(\cdot)$, it follows that

$$W_0 = \underset{W \in L^2(0,1)}{\arg \min} \, I_0(W) = \frac{V^{in} g_l^i + V^{ex} g_l^e}{g_l^i + g_l^e}. \tag{3.15}$$

In Fig. 3.3 we present a numerical simulation, for $N^i = N^e = 255$. Also $\varepsilon = 10^{-2}$, $V^{in} = -10$, $V^{ex} = 65$, $g_l^i = 4 \times 10^{-2}$, $g_l^e = 10^{-2}$. The synapses are disposed periodically, in an alternate fashion. Again, the exact solution is plotted in a solid line. The values of the MsFEM solution are indicated by $\square$, and the nodal values of the classical Galerkin solution are plotted with $*$.

Fig. 3.3 Numerical test in the case of large number of synapses. The exact solution (*solid*) and the MsFEM solution (□) agree. The solution by classical finite element method (∗) delivers a good approximation in the interior of the domain, but not close to the boundaries

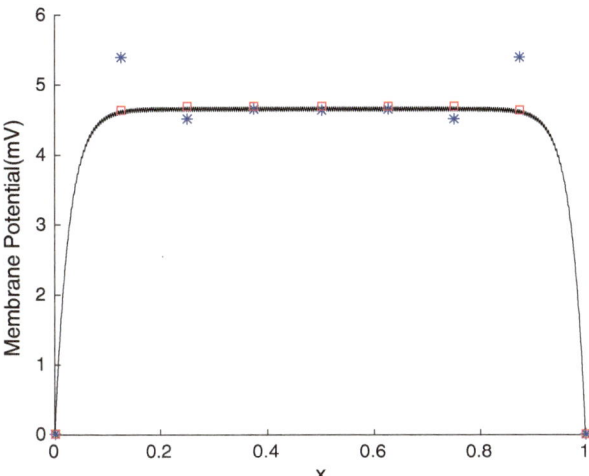

3.5   Conclusions

Sometimes one-dimensional equations are more than just an over-simplification designed to further our understanding of a problem, but indeed a verisimilar model of a rather complex phenomenon [47, 177, 185]. That is the case in neuroscience when there is interest in the propagation of pulses along a neuron.

The whole neuronal model for pulse propagation is a nonlinear time dependent system of PDEs, but we consider here a rather simplified equation that is linear and elliptic. It is of reaction-diffusion type, but the presence of deltas causes the onset of internal layers. We show that this is not a problem when the MsFEM is considered since the basis functions adapt to the problem data, resulting in a robust method. We discussed some different asymptotic limits and illustrate the theoretical results with some numerics.

Chapter 4
Two-Dimensional Reaction-Diffusion Equations

Abstract We now investigate two-dimensional domains, and consider a reaction-diffusion problem. We first develop an asymptotic expansion for the solution, and this time we show how to deal with the boundary layer in a two-dimensional problem, assuming that the boundary is smooth. We then derive an estimate for non-smooth domains. Finally, we present a numerical scheme that is a variation of the Residual Free Bubble method that works well for the problem under consideration.

4.1 Introduction

One-dimensional second order problems with reaction terms are considered in Chap. 2, and here we extend the problem to two-dimensional domains. The main difference relies on how to deal with the boundary correctors. For smooth domains, it is possible to use *boundary-fitted coordinates* [19], a local diffeomorphism that permits the description of the boundary correctors in a quite explicit form. For more general domains this technique fails, and we estimate only how close the first term of the asymptotic expansion is to the exact solution; see also [64, 147, 150].

We also propose a Petrov–Galerkin finite element method based on space enrichment that is very similar to the Residual Free Bubbles method but also inherits some aspects of the Multiscale Finite Element Method [101, 102, 134, 136, 137].

In the next section we consider the asymptotic expansion for the reaction-diffusion equation in both smooth and non-smooth cases, and then, in Sect. 4.3, consider an enriched finite element method. We conclude the chapter with some final remarks.

4.2 The Continuous Problem

In what follows next, we develop an asymptotic expansion for the solution of a reaction-diffusion problem, paying special attention to the boundary layer terms. Since it is a two-dimensional domain, it requires some special techniques that differ considerably from what was presented before. Some error estimates for the smooth and non-smooth boundaries are then considered as well.

© The Author(s) 2017

A.L. Madureira, *Numerical Methods and Analysis of Multiscale Problems*,
SpringerBriefs in Mathematics, DOI 10.1007/978-3-319-50866-5_4

4.2.1 Asymptotic Expansion

Consider the singular perturbed reaction-diffusion problem

$$-\varepsilon^2 \Delta u + u = f \quad \text{in } \Omega,$$
$$u = 0 \quad \text{on } \partial\Omega, \tag{4.1}$$

where Ω is a smooth two-dimensional bounded domain and ε is a positive constant. Also assume that f is smooth.

Consider the series

$$u^0 + \varepsilon^2 u^2 + \varepsilon^4 u^4 + \cdots$$

and formally substitute it in (4.1). Then

$$u^0 + \varepsilon^2 \big(-\Delta u^0 + u^2\big) + \cdots + \varepsilon^{2i} \big(-\Delta u^{2i-2} + u^{2i}\big) + \cdots = f.$$

By comparing the different powers of ε, it is natural to require that

$$u^0 = f, \quad u^2 = \Delta u^0, \ldots, \quad u^{2i} = \Delta u^{2i-2}, \ldots. \tag{4.2}$$

Since the u^i are already well defined, we cannot impose the zero Dirichlet boundary condition. We again correct this by introducing boundary correctors. We would like to have

$$-\varepsilon^2 \Delta U + U = 0, \quad \text{in } \Omega, \qquad U = u^0 + \varepsilon^2 u^2 + \varepsilon^4 u^4 + \cdots \quad \text{on } \partial\Omega, \tag{4.3}$$

and formally expand

$$U \sim U^0 + \varepsilon U^1 + \varepsilon^2 U^2 + \cdots. \tag{4.4}$$

Motivated by the one-dimensional problem, we expect the boundary correctors to have only a "local" influence, and we introduce for that purpose boundary-fitted coordinates. We digress now to introduce these coordinates [19, 50].

Suppose that $\partial\Omega$ is arc-length parametrized in counter-clockwise direction by $\mathbf{z}(\theta) = (X(\theta), Y(\theta))$. Let $\mathbf{s} = (X', Y')$, $\mathbf{n} = (Y', -X')$ denote the tangent and the outward normal of $\partial\Omega$, and define the subdomain $\Omega_b \subset \Omega$,

$$\Omega_b = \big\{\mathbf{z} - \rho\mathbf{n} : \mathbf{z} \in \partial\Omega, \, 0 < \rho < \rho_0\big\}, \tag{4.5}$$

where ρ_0 is a positive number smaller than the minimum radius of curvature of $\partial\Omega$. With L denoting the arc-length of $\partial\Omega$, then

$$\mathbf{x} : (0, \rho_0) \times \mathbb{R}/L \to \Omega_b,$$

Fig. 4.1 Boundary-fitted
coordinates

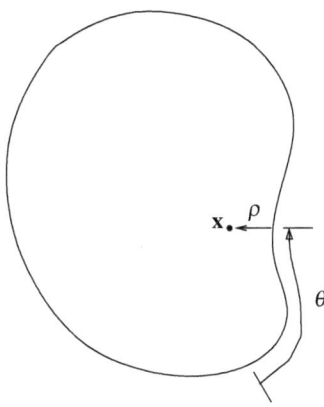

where

$$\mathbf{x}(\rho, \theta) = \mathbf{z}(\theta) - \rho \mathbf{n}(\theta),$$

is a diffeomorphism. See Fig. 4.1.

Remark 4.1 The smoothness of the domain is important here. In particular for a polygon, the above transformation is not a diffeomorphism.

From planar differential geometry [70, 198], the curvature κ of $\partial \Omega$ can be defined through the formula $\mathbf{z}'' = -\kappa \mathbf{n}$, and then

$$\kappa(\theta) = -\mathbf{z}'' \cdot \mathbf{n} = -Y'X'' + X'Y''. \tag{4.6}$$

Note then that

$$\frac{\partial \mathbf{x}}{\partial \rho} = -\mathbf{n}(\theta) = (-Y', X'), \qquad \frac{\partial \mathbf{x}}{\partial \theta} = \mathbf{z}'(\theta) - \rho \mathbf{n}'(\theta) = (X' - \rho Y'', Y' + \rho X''),$$

and thus

$$\nabla_{(\rho, \theta)} \mathbf{x} = \begin{pmatrix} -Y' & X' - \rho Y'' \\ X' & Y' + \rho X'' \end{pmatrix}.$$

Using (4.6), and $\rho < \rho_0$, we obtain that $\det \nabla_{(\rho, \theta)} \mathbf{x} = -1 + \rho \kappa \neq 0$ in Ω_b, see definition (4.5). Inverting the above matrix, we have that

$$\nabla_\mathbf{x} \begin{pmatrix} \rho & \theta \end{pmatrix} = \frac{1}{J} \begin{pmatrix} Y' + \rho X'' & -X' + \rho Y'' \\ -X' & -Y' \end{pmatrix},$$

where $J(\rho, \theta) = -1 + \rho \kappa(\theta)$. Hence,

$$\nabla_\mathbf{x} \rho = \frac{1}{J} \mathbf{n} + \frac{\rho}{J} \mathbf{z}'' = \frac{1}{J}(1 - \rho \kappa)\mathbf{n} = -\mathbf{n}, \qquad \nabla_\mathbf{x} \theta = -\frac{1}{J} \mathbf{s}. \tag{4.7}$$

Finally, the change of coordinates yields

$$\frac{\partial f}{\partial \alpha} = \frac{\partial f}{\partial \theta}\frac{\partial \theta}{\partial \alpha} + \frac{\partial f}{\partial \rho}\frac{\partial \rho}{\partial \alpha}, \qquad \text{for } \alpha = 1, 2,$$

for an arbitrary function f. So, from the product rule and applying the above formula,

$$\begin{aligned}
\frac{\partial^2 U}{\partial \alpha^2} &= \frac{\partial}{\partial \alpha}\left(\frac{\partial U}{\partial \theta}\right)\frac{\partial \theta}{\partial \alpha} + \frac{\partial U}{\partial \theta}\frac{\partial^2 \theta}{\partial \alpha^2} + \frac{\partial}{\partial \alpha}\left(\frac{\partial U}{\partial \rho}\right)\frac{\partial \rho}{\partial \alpha} + \frac{\partial U}{\partial \rho}\frac{\partial^2 \rho}{\partial \alpha^2} \\
&= \frac{\partial^2 U}{\partial \theta^2}\frac{\partial \theta^2}{\partial \alpha} + 2\frac{\partial^2 U}{\partial \rho \partial \theta}\frac{\partial \theta}{\partial \alpha}\frac{\partial \rho}{\partial \alpha} + \frac{\partial U}{\partial \theta}\frac{\partial^2 \theta}{\partial \alpha^2} + \frac{\partial^2 U}{\partial \rho^2}\frac{\partial \rho^2}{\partial \alpha} + \frac{\partial U}{\partial \rho}\frac{\partial^2 \rho}{\partial \alpha^2}.
\end{aligned}$$

Adding in $\alpha = 1, 2$,

$$\Delta U = \frac{\partial^2 U}{\partial \theta^2}|\nabla \theta|^2 + 2\frac{\partial^2 U}{\partial \rho \partial \theta}\nabla \rho \cdot \nabla \theta + \frac{\partial U}{\partial \theta}\Delta \theta + \frac{\partial U}{\partial \rho}\Delta \rho + \frac{\partial^2 U}{\partial \rho^2}|\nabla \rho|^2,$$

and using that (show it)

$$\Delta \theta = \frac{\rho \kappa'}{J^3}, \qquad \Delta \rho = -\frac{\kappa}{J}, \tag{4.8}$$

the expression for the Laplacian in these new coordinates follows from (4.7):

$$\begin{aligned}
\Delta U &= \frac{\partial^2 U}{\partial \rho^2} - \frac{\kappa}{J}\frac{\partial U}{\partial \rho} + \frac{1}{J^2}\frac{\partial^2 U}{\partial \theta^2} + \frac{\rho \kappa'}{J^3}\frac{\partial U}{\partial \theta} \\
&= \frac{\partial^2 U}{\partial \rho^2} + \sum_{j=0}^{\infty} \rho^j\left(d_1^j\frac{\partial U}{\partial \rho} + d_2^j\frac{\partial^2 U}{\partial \theta^2} + d_3^j\frac{\partial U}{\partial \theta}\right),
\end{aligned} \tag{4.9}$$

where we formally replace each coefficient with its respective Taylor expansion,

$$d_1^j = -[\kappa(\theta)]^{j+1}, \quad d_2^j = (j+1)[\kappa(\theta)]^j, \quad d_3^j = \frac{j(j+1)}{2}[\kappa(\theta)]^{j-1}\kappa'(\theta).$$

Defining the new variable $\hat{\rho} = \varepsilon^{-1}\rho$ and using the same name for functions different only up to this change of coordinates, we have from (4.9) that

$$\Delta U = \varepsilon^{-2}\partial_{\hat{\rho}\hat{\rho}}U + \sum_{j=0}^{\infty}(\varepsilon\hat{\rho})^j\left(d_1^j\varepsilon^{-1}\partial_{\hat{\rho}}U + d_2^j\partial_{\theta\theta}U + d_3^j\partial_{\theta}U\right). \tag{4.10}$$

Aiming to solve (4.3) we formally use (4.4) and (4.10), collect together terms with same order of ε and for $k \geq 2$, pose the following sequence of problems parametrized by θ:

$$-\partial_{\hat{\rho}\hat{\rho}} U^k + U^k = F_k \qquad \text{in } \mathbb{R}^+, \tag{4.11}$$

$$U^k(0,\theta) = u^k(0,\theta),$$

with the convention that $u^k = 0$ for k odd, and

$$F_0 = 0, \qquad F_1 = a_1^0 \partial_{\hat{\rho}} U^0,$$

$$F_k = \sum_{j=0}^{k-1} \hat{\rho}^j a_1^j \partial_{\hat{\rho}} U^{k-j-1} + \sum_{j=0}^{k-2} \hat{\rho}^j \left(a_2^j \partial_{\theta\theta} U^{k-j-2} + a_3^j \partial_\theta U^{k-j-2} \right), \qquad \text{for } k \geq 2.$$

With the boundary layer terms defined, we gather that the asymptotic expansion is given by

$$u^\epsilon(\mathbf{x}) \sim u^0(\mathbf{x}) + \varepsilon^2 u^(\mathbf{x}) + \varepsilon^4 u^4(\mathbf{x}) + \cdots$$

$$- \chi(\rho)\left[U^0(\varepsilon^{-1}\rho, \theta) + \varepsilon U^1(\varepsilon^{-1}\rho, \theta) + \varepsilon^2 U^2(\varepsilon^{-1}\rho, \theta) + \cdots \right], \tag{4.12}$$

where χ is a smooth cut-off function identically one if $0 \leq \rho \leq \rho_0/3$ and identically zero if $\rho \geq 2\rho_0/3$.

Remark 4.2 The presence of a cut-off function is essential, since the boundary-fitted coordinates are only defined in a neighborhood of $\partial\Omega$. Since the boundary correctors decay exponentially to zero, the cut-off functions will introduce only a exponentially small, thus negligible, error.

4.2.2   Error Estimates for the Asymptotic Expansion

Here we estimate how close a truncated asymptotic expansion approximates the exact solution. We assume that c is a constant that might depend on the domain, and the right-hand side f.

We first estimate the boundary correctors. For every s and positive integer i, there exist ε-independent constants c and α such that

$$\|u^{2i}\|_{H^s(\Omega)} \leq c, \qquad U^i(\hat{\rho}) \leq c \exp(-\alpha\hat{\rho}). \tag{4.13}$$

Indeed,

$$\|u^{2i}\|_{H^s(\Omega)} \leq c\|\Delta^i f\|_{H^s(\Omega)} \leq c\|f\|_{H^{s+2i}(\Omega)}.$$

The decay of U^i follows from standard ODE results and induction in i. From (4.13) we have that

$$\|U^i\|_{L^2(\Omega)} + \left\|\frac{\partial U^i}{\partial \theta}\right\|_{L^2(\Omega)} + \varepsilon\|U^i\|_{H^1(\Omega)} \leq c\varepsilon^{1/2}.$$

Although (4.12) is a formal expansion, a rigorous error estimate shows that the difference between the exact solution and a truncated asymptotic expansion is of the same order of the first term omitted in the expansion. In fact, define

$$e_{2N}(\mathbf{x}) = u^\varepsilon(\mathbf{x}) - \sum_{k=0}^{N} \varepsilon^{2k} u^{2k}(\mathbf{x}) + \chi(\rho) \sum_{k=0}^{2N} \varepsilon^k U^k(\varepsilon^{-1}\rho, \theta). \tag{4.14}$$

In the theorem below we bound the $H^1(\Omega)$ norm of e_{2N}.

Theorem 4.1 *For any positive integer N, there exists a constant c such that the difference between the truncated asymptotic expansion and the original solution measured in the original domain is bounded as follows:*

$$\|e_{2N}\|_{H^1(\Omega)} \leq c\varepsilon^{2N+1/2}. \tag{4.15}$$

Before we prove Theorem 4.1, we develop some other estimates. For instance, in the $L^2(\Omega)$ norm, we have from the triangle inequality that

$$\|e_{2N}\|_{L^2(\Omega)} \leq \|e_{2N+2}\|_{H^1(\Omega)} + \|e_{2N+2} - e_{2N}\|_{L^2(\Omega)}.$$

Since

$$(e_{2N+2} - e_{2N})(\mathbf{x}) = -\varepsilon^{2N+1} \chi(\rho) U^{2N+1}(\varepsilon^{-1}\rho, \theta)$$
$$+ \varepsilon^{2N+2}\left[u^{2N+2}(\mathbf{x}) - \chi(\rho) U^{2N+2}(\varepsilon^{-1}\rho, \theta)\right], \tag{4.16}$$

we conclude that

$$\|e_{2N}\|_{L^2(\Omega)} \leq c\varepsilon^{2N+3/2},$$

for N non-negative.

Using similar arguments, it is possible to compute *interior estimates*, which achieve better convergence in regions "far away" from the lateral boundary of the plate. The reason for the improvement in such subdomains is that the influence of the boundary layer is negligible. Table 4.1 presents these interior and various other error estimates. We assume that f is a sufficiently smooth function and we show only the order of the norms with respect to ε. "BL" stands for "Boundary Layer" and the "Relative Error" column presents the norm of e_{2N} divided by the norm of u^ε. In parentheses are the interior estimates, when these are better than the global estimates.

The remainder of this section contains a proof of Theorem 4.1.

Table 4.1 Order with respect to ε of the exact solution, the first term of the boundary layer expansion, and the difference between the solution and a truncated asymptotic expansion in various norms

Norm	u^ε	BL	$e_{2N}, N \geq 0$	Relative error
$\|\cdot\|_{L^2(\Omega)}$	1	$\varepsilon^{1/2}$	$\varepsilon^{2N+3/2}(\varepsilon^{2N+2})$	$\varepsilon^{2N+3/2}(\varepsilon^{2N+2})$
$\|\partial_\rho \cdot\|_{L^2(\Omega)}$	$\varepsilon^{-1/2}(1)$	$\varepsilon^{-1/2}$	$\varepsilon^{2N+1/2}(\varepsilon^{2N+2})$	$\varepsilon^{2N+1}(\varepsilon^{2N+2})$
$\|\partial_\theta \cdot\|_{L^2(\Omega)}$	1	$\varepsilon^{1/2}$	$\varepsilon^{2N+3/2}(\varepsilon^{2N+2})$	$\varepsilon^{2N+3/2}(\varepsilon^{2N+2})$
$\|\cdot\|_{H^1(\Omega)}$	$\varepsilon^{-1/2}(1)$	$\varepsilon^{-1/2}$	$\varepsilon^{2N+1/2}(\varepsilon^{2N+2})$	$\varepsilon^{2N+1}(\varepsilon^{2N+2})$

Interior estimates are also presented, in parentheses, whenever they are better than the global estimates

Definition 4.1 Set

$$u_{2N}(\mathbf{x}) = \sum_{k=0}^{N} \varepsilon^{2k} u^{2k}(\mathbf{x}), \quad U_{2N}(\mathbf{x}) = \sum_{k=0}^{2N} \varepsilon^k U^k(\varepsilon^{-1}\rho, \theta, x_3).$$

Some results regarding the boundary layer terms are collected below.

Lemma 4.1 *For any positive integer N, there exist positive constants C and α such that*

$$\|\chi' U_{2N}\|_{L^2(\Omega)} + \varepsilon \|\chi' \partial_\rho U_{2N}\|_{L^2(\Omega)} \leq C \exp(-\alpha \varepsilon^{-1}). \tag{4.17}$$

Also, for all $v \in H_0^1(\Omega)$,

$$\left| \int_\Omega \nabla U_{2N} \cdot \nabla(\chi v) + U_{2N}\chi v \, d\mathbf{x} \right| \leq C\varepsilon^{2N} \|v\|_{H^1(\Omega)}. \tag{4.18}$$

Proof The inequalities (4.17) follow from a change of coordinates, (4.13), and the definition of χ. To see that (4.18) holds, first rewrite (4.9) as a finite series, using Taylor expansion with remainders. Then the result follows from the definition of U_{2N}, (4.11), and (4.13). □

Proof (Theorem 4.1) Let $v \in H_0^1(\Omega)$. If we define

$$E(2N, v) = \int_\Omega \varepsilon \nabla(u^\varepsilon - u_{2N}) \nabla v + (u^\varepsilon - u_{2N})v \, d\mathbf{x},$$

then, by construction of the asymptotic expansion, we have

$$E(2N, v) = \int_\Omega fv \, d\mathbf{x} - \sum_{k=0}^{N} \varepsilon^{2k} \int_\Omega \left(\varepsilon^2 \nabla u^{2k} \nabla v + u^{2k}v \right) d\mathbf{x}$$

$$= -\varepsilon^{2N+2} \int_\Omega \nabla u^{2N} \nabla v \, d\mathbf{x},$$

and we conclude from (4.13) that

$$|E(2N, v)| \le c\varepsilon^{2N+2}\|v\|_{H^1(\Omega)}. \tag{4.19}$$

We also have

$$\left|\int_\Omega \nabla(\chi U_{2N}) \nabla v - \nabla U_{2N} \nabla(\chi v)\, d\mathbf{x}\right|$$

$$\le \left(\|\chi' U_{2N}\|_{L^2(\Omega)} + \|\chi' \nabla U_{2N}\|_{L^2(\Omega)}\right)\|v\|_{H^1(\Omega)} \le c\exp(-\alpha\varepsilon^{-1})$$

by Lemma 4.1. Hence

$$\left|\int_\Omega \left[\nabla(\chi U_{2N}) \nabla v + \chi U_{2N} v\right] d\mathbf{x}\right|$$

$$\le \left|\int_\Omega \left[\nabla U_{2N} \nabla(\chi v) + U_{2N}\chi v\right] d\mathbf{x}\right| + c\exp(-\alpha\varepsilon^{-1}) \le C\varepsilon^{2N}\|v\|_{H^1(\Omega)}. \tag{4.20}$$

Finally, since e_{2N} vanishes on $\partial\Omega$,

$$\|e_{2N}\|_{H^1(\Omega)}^2 = \int_\Omega |\nabla e_{2N}|^2 + (e_{2N})^2\, d\mathbf{x}$$

$$= E(2N, e_{2N}) + \int_\Omega \left[\nabla(\chi U_{2N}) \nabla e_{2N} + \chi U_{2N} e_{2N}\right] d\mathbf{x}$$

$$\le C\varepsilon^{2N}\|e_{2N}\|_{H^1(\Omega)}, \tag{4.21}$$

from (4.19) and (4.20). The estimate (4.21) is not sharp yet, so we use the triangle inequality:

$$\|e_{2N}^\epsilon\|_{H^1(\Omega)} \le \|e_{2N+2} - e_{2N}\|_{H^1(\Omega)} + c\varepsilon^{2N+2},$$

and then the result follows from (4.16). $\square$

4.2.3 Estimates for Non-smooth Domain

We consider now estimates for problem (4.1) when the domain Ω is not necessarily smooth, for example, if Ω is polygonal.

We shall need the following interpolation inequality:

$$\|g\|_{s+v,\Omega}^u \le \|g\|_{s,\Omega}^{u-v}\|g\|_{s+u,\Omega}^v, \quad s \ge 0,\ u \ge v \ge 0. \tag{4.22}$$

Furthermore, for $g \in L^2(\Omega)$, let $\Delta^{-1}g$ be the unique function in $H^2(\Omega) \cap H_0^1(\Omega)$ whose Laplacian is equal to g. Then

$$C^{-1}\|\Delta^{-1}g\|_{s+2,\Omega} \leq \|g\|_{s,\Omega} \leq C\|\Delta^{-1}g\|_{s+2,\Omega}, \quad s \geq 0. \tag{4.23}$$

See [19] for further details.

From [7], we have the following result. We reproduce here the proof, in some detail.

Lemma 4.2 *Let $f \in H^1(\Omega)$, and u be the solution of (4.1). Then there exists a constant c that might depend on Ω such that*

$$\varepsilon^2\|\nabla u\|_{0,\Omega}^2 + \|u - f\|_{0,\Omega}^2 \leq c\big(\varepsilon\|f\|_{0,\partial\Omega}^2 + \varepsilon^2\|f\|_{1,\Omega}^2\big).$$

Proof Multiplying the differential equation by $-\Delta u$, and integrating by parts yields

$$\varepsilon^2\|\Delta u\|_{0,\Omega}^2 + \|\nabla u\|_{0,\Omega}^2 = \int_\Omega \nabla f \nabla u \, d\mathbf{x} - \int_{\partial\Omega} f \frac{\partial u}{\partial n} \, ds.$$

Note that using the trace inequality, and (4.22) with $u = 1$, $v = 1/2$, and $s = 1$, we find that

$$\left\|\frac{\partial u}{\partial n}\right\|_{0,\partial\Omega} \leq \|u\|_{3/2,\Omega} \leq \|u\|_{1,\Omega}^{1/2}\|u\|_{2,\Omega}^{1/2}.$$

Hence,

$$\left\|\frac{\partial u}{\partial n}\right\|_{0,\partial\Omega}^2 \leq c\big(\varepsilon^{-1}\|u\|_{1,\Omega}^2 + \varepsilon\|u\|_{2,\Omega}^2\big).$$

So, for any $\delta_1 > 0$,

$$\left|\int_{\partial\Omega} f \frac{\partial u}{\partial n} \, ds\right| \leq \|f\|_{0,\partial\Omega}\left\|\frac{\partial u}{\partial n}\right\|_{0,\partial\Omega} \leq c_{\delta_1}\varepsilon^{-1}\|f\|_{0,\partial\Omega}^2 + \delta_1\varepsilon\left\|\frac{\partial u}{\partial n}\right\|_{0,\partial\Omega}^2$$

$$\leq c_{\delta_1}\varepsilon^{-1}\|f\|_{0,\partial\Omega}^2 + c\delta_1\big(\|u\|_{1,\Omega}^2 + \varepsilon^2\|u\|_{2,\Omega}^2\big)$$

$$\leq c_{\delta_1}\varepsilon^{-1}\|f\|_{0,\partial\Omega}^2 + c\delta_1\big(\|u\|_{1,\Omega}^2 + \varepsilon^2\|\Delta u\|_{0,\Omega}^2\big),$$

where the norm equivalence (4.23) was used in the last inequality above. Similarly, for any $\delta_2 > 0$,

$$\left|\int_\Omega \nabla f \nabla u \, d\mathbf{x}\right| \leq \|f\|_{1,\Omega}\|u\|_{1,\Omega} \leq c_{\delta_2}\|f\|_{1,\Omega}^2 + \delta_2\|u\|_{1,\Omega}^2.$$

It follows from these estimates, and careful choices of δ_1 and δ_2 that

$$\varepsilon^2\|\Delta u\|_{0,\Omega}^2 + \|\nabla u\|_{0,\Omega}^2 \leq c\big(\varepsilon^{-1}\|f\|_{0,\partial\Omega}^2 + \|f\|_{1,\Omega}^2\big).$$

We finally multiply the above inequality by ε^2, and use that

$$\varepsilon^2 \Delta u = u - f$$

to conclude the proof.   □

   Hence, if $f \in H^1(\Omega)$, the solution u converges to f in $L^2(\Omega)$ as $\varepsilon \to 0$.

4.3 Finite Element Approximations

Here we present an introductory discussion on how to use finite element techniques to approximate the singular perturbed reaction-diffusion problem (4.1), related to the differential operator

$$\mathscr{L} u = -\varepsilon^2 \Delta u + u. \tag{4.24}$$

   The reaction dominated regime of problem (4.1) possesses boundary layers that challenge good numerical approximations. Some previous works, e.g. [96], tried to tame such difficulty using stabilized methods, see Sect. 1.3. As pointed out previously, such methods are partially justified by its relation with bubble enriched methods [28, 39, 40].
   In what follows, we consider a partition $\mathscr{T}_h = \{K_j\}$ of Ω into "quadrilateral elements." Let $Q^1(\Omega)$ be the space of continuous functions in Ω that are bilinear polynomials in each quadrilateral, and define $Q_0^1(\Omega) = Q^1(\Omega) \cap H_0^1(\Omega)$.
   We propose a scheme that has good approximation properties for all positive ε.

4.3.1 General Comments

Classical methods fail to deliver a reasonable approximation to problem (4.1) under the reaction dominated regime. The reason is the lack of coercivity—see Chap. 2 for the one-dimensional analysis of a similar problem.
   The Residual Free Bubble (RFB) Method yields better results, but is still not satisfying. The reason is that the bubbles introduce boundary layers that do not correspond to the ones present in the asymptotic of the exact solution. Aiming to correct that, but still keeping the good properties of the Residual Free Bubbles, we propose a Petrov–Galerkin method [102]. In this method, the test space is enriched with bubbles, as in the RFB. Then some of the problems are local, allowing an efficient static condensation. The trial space is derived in an ad hoc form; since there is freedom to impose local boundary conditions for the local problems, we use a "restriction" of the differential operator over the edges, similar actually to (4.11). This originates ODEs over the edges that can be solved a priori. Similar ideas

appear in [136, 137] for PDEs with oscillatory coefficients and were tested with good results, but apparently never analyzed. See [101] for an analysis of the present method.

4.3.2 Enriching Finite Element Spaces

In this section we describe the work developed in [101, 102]. We are interested in finding a finite element discretization for (4.1) that is stable and coarse mesh accurate for all ε. We use the approach of enriching the finite element space. The idea is to add special functions to the usual polynomial spaces to stabilize and improve accuracy of the Galerkin method. This goes along the philosophy of the *Residual Free Bubbles (RFB)* method [41, 42, 107–109] briefly described in Sect. 1.3, and actually extends it.

4.3.2.1 New Enriched Choice

Consider

$$a(u^h, v^h) = \langle f, v^h \rangle \quad \text{for all } v^h \in Q_0^1(\Omega),$$

where

$$a(u^h, v^h) = \varepsilon^2 \int_\Omega \nabla u^h \cdot \nabla v^h \, d\mathbf{x} + \int_\Omega u^h v^h \, d\mathbf{x}, \qquad \langle f, v^h \rangle = \int_\Omega f v \, d\mathbf{x},$$

and

$$U^h = Q_0^1(\Omega) \oplus E^*(\Omega),$$

as the trial space, where $E^*(\Omega)$ is yet to be defined. As the test space, we set

$$Q_0^1(\Omega) \oplus B,$$

where the bubble space B is defined in (1.18).

In our Petrov–Galerkin formulation, we seek $u^h = u^1 + u^* \in U^h$, where $u^1 \in Q_0^1(\Omega)$ and $u^* \in E^*(\Omega)$, and

$$a(u^h, v^h) = \langle f, v^h \rangle \quad \text{for all } v^h \in Q_0^1(\Omega), \tag{4.25}$$

$$a(u^h, v) = \langle f, v^h \rangle \quad \text{for all } v \in H_0^1(K) \text{ and all } K \in \mathscr{T}_h. \tag{4.26}$$

It follows from (4.26) that, for every $K \in \mathscr{T}_h$,

$$\mathscr{L} u^* = f - \mathscr{L} u^1 \quad \text{in } K. \tag{4.27}$$

Note that we still need to impose boundary conditions on ∂K to define u^*. The usual residual-free bubble formulation subjects u^* to a homogeneous Dirichlet element boundary condition, i.e., $u^* = 0$ on ∂K, for all elements K. Herein, we replace this condition by a more sophisticated choice, based on ideas by Hou and Wu [136].

To determine u^* uniquely, we impose the boundary conditions

$$u^* = 0 \quad \text{on } \partial K, \text{ if } \partial K \subset \partial \Omega,$$

$$\mathscr{L}_{\partial K} u^* = \mathscr{R}(f - \mathscr{L} u^1) \quad \text{on } \partial K, \text{ if } \partial K \not\subset \partial \Omega, \tag{4.28}$$

$$u^* = 0 \quad \text{on all vertices of } K, \tag{4.29}$$

where $\mathscr{R}$ is the trace operator, and we choose

$$\mathscr{L}_{\partial K} v = -\varepsilon^2 \partial_{ss} v + v, \tag{4.30}$$

where s denotes a variable that runs along ∂K. Note that the restriction of f to K must be regular enough so that its trace on ∂K makes sense. Henceforth, we assume that $f \in Q^1(\Omega)$.

The choice of (4.30) is ad hoc, and by no means unique. But it can be justified under the light of asymptotic analysis. Indeed this is the equation satisfied by the boundary correctors, *in the direction of the boundary layers*, see (4.11). Hence, we are enriching the space of polynomials with functions that have the same behavior as the correctors. In some sense, the polynomial part of the approximation (u^1 in our case) "captures" the smooth behavior of the exact solution. The local, "multiscale behavior" is seen by the enrichment functions (u^* in our case) that adds its contribution to the final formulation, without making the method expensive. In other words, it is possible to describe the multiscale characteristics of a solution for a singular perturbed PDE, without having to resolve all the fine scales with a refined mesh.

We can formally write the solution of (4.27)–(4.30) as

$$u^* = \mathscr{L}_*^{-1}(f - \mathscr{L}_{\mathscr{T}} u^1) \in L^2(\Omega), \quad \text{where } \mathscr{L}_{\mathscr{T}} = \sum_{K \in \mathscr{T}_h} \chi_K \mathscr{L}, \tag{4.31}$$

and χ_K is the characteristic function of K. We finally set $E^*(\Omega) = \mathscr{L}_*^{-1} Q^1(\Omega)$.

Substituting (4.31) in (4.25), we gather that

$$a((I - \mathscr{L}_*^{-1} \mathscr{L}_{\mathscr{T}})u^1, v^h) = \langle f, v^h \rangle - a(\mathscr{L}_*^{-1} f, v^h) \quad \text{for all } v^h \in Q_0^1(\Omega). \tag{4.32}$$

Finally, $u^h = (I - \mathscr{L}_*^{-1} \mathscr{L}_{\mathscr{T}})u^1 + \mathscr{L}_*^{-1} f$. Note nevertheless that, because of (4.29), $u^h = u^1$ at the nodal points, as in the usual polynomial Galerkin formulation.

Remark 4.3 Note that our particular choice of test space allowed the *static condensation* procedure, i.e., we were able to write u^* with respect to u^1 and f, as in (4.31).

The matrix formulation can be obtained as follows. Under the assumption that $f \in Q^1(\Omega)$, we write

$$f = \sum_{j \in J} f_j \psi_j, \quad u^1 = \sum_{j \in J_0} u_j^1 \psi_j$$

where J and J_0 are the set of indexes of total and interior nodal points, $\{\psi_j\}_{j \in J}$ form a basis of $Q^1(\Omega)$, and $\{\psi_j\}_{j \in J_0}$ form a basis of $Q_0^1(\Omega)$. Substituting in (4.31), we have that

$$u^* = \sum_{j \in J} f_j \mathcal{L}_*^{-1} \mathcal{L}_{\mathscr{T}} \psi_j - \sum_{j \in J_0} u_j^1 \mathcal{L}_*^{-1} \mathcal{L}_{\mathscr{T}} \psi_j, \tag{4.33}$$

where we used that

$$\mathcal{L}_{\mathscr{T}} \psi_j = \psi_j. \tag{4.34}$$

To write the variational formulation in an explicit form, it is convenient to define $\lambda_j = (I - \mathcal{L}_*^{-1} \mathcal{L}_{\mathscr{T}}) \psi_j$. Hence, (4.32) reads as

$$\sum_{j \in J_0} a(\lambda_j, \psi_i) u_j^1 = \sum_{j \in J} \left[\langle \psi_j, \psi_i \rangle - a(\mathcal{L}_*^{-1} \psi_j, \psi_i) \right] f_j \quad \text{for all } i \in J_0. \tag{4.35}$$

Using the definition of the bilinear form $a(\cdot, \cdot)$, and (4.34), yields

$$\sum_{j \in J_0} a(\lambda_j, \psi_i) u_j^1 = \sum_{j \in J} \int_{\Omega} \left[\lambda_j \psi_i + \varepsilon^2 \nabla(\lambda_j - \psi_j) \cdot \nabla \psi_i \right] d\mathbf{x} f_j \quad \text{for all } i \in J_0. \tag{4.36}$$

Concrete computations of the matrix formulation follows.

4.3.2.2 Solving Local Problems

A core and troublesome issue in the present method is solving the local problems. From its definition, λ_j solves

$$\mathcal{L} \lambda_j = 0 \quad \text{in } K,$$

$$\mathcal{L}_{\partial K} \lambda_j = 0 \quad \text{on } \partial K, \qquad \lambda_j = \begin{cases} 1 & \text{on the } j\text{th vertices of } \mathscr{T}_h, \\ 0 & \text{on the other vertices of } \mathscr{T}_h, \end{cases} \tag{4.37}$$

Remark 4.4 In fact, we have that $\mathcal{L}_{\partial K} \lambda_j = \mathcal{L}_{\partial K} \psi_j - \mathcal{L} \psi_j$ on ∂K. Since we are assuming that ψ_j is bilinear over a rectangular mesh, we have that ψ_j is still linear over ∂K. Hence, $\mathcal{L}_{\partial K} \psi_j = \psi_j$.

Consider now a rectangular straight mesh. Our goal is to find λ_j. Without loss of generality, consider a rectangle K with vertices $1, \ldots, 4$ at $(0,0)$, $(h_x, 0)$, (h_x, h_y), and $(0, h_y)$. So, again without loss of generalization, we want to find λ_1. We have that

$$-\varepsilon^2 \Delta \lambda_1 + \lambda_1 = 0 \quad \text{in } K. \tag{4.38}$$

On the side $y = 0$, we have that

$$-\varepsilon^2 \partial_{xx} \lambda_1 + \lambda_1 = 0 \quad \text{for } x \text{ in } (0, 1),$$

$$\lambda_1(0,0) = 1, \quad \lambda_1(h_x, 0) = 0.$$

Hence,

$$\lambda_1(x, 0) = \mu_x(x) := -\frac{\sinh\left(\varepsilon^{-1}(x - h_x)\right)}{\sinh\left(\varepsilon^{-1} h_x\right)}. \tag{4.39}$$

Similarly,

$$\lambda_1(0, y) = \mu_y(y) := -\frac{\sinh\left(\varepsilon^{-1}(y - h_y)\right)}{\sinh\left(\varepsilon^{-1} h_y\right)}, \qquad \lambda_1(h_x, y) = \lambda_1(x, h_y) = 0. \tag{4.40}$$

We propose two simple closed forms for λ_1, none of which satisfy (4.38)–(4.40) exactly. If we set $\lambda_1(x, y) = \mu_x(x)\mu_y(y)$, then (4.39)–(4.40) hold, but

$$-\varepsilon^2 \Delta \lambda_1 + 2\lambda_1 = 0 \quad \text{in } K,$$

thus (4.38) is *not* satisfied.

If we let

$$\lambda_1(x, y) = \frac{\sinh\left(\varepsilon^{-1}\sqrt{\tfrac{1}{2}}(x - h_x)\right) \sinh\left(\varepsilon^{-1}\sqrt{\tfrac{1}{2}}(y - h_y)\right)}{\sinh\left(\varepsilon^{-1}\sqrt{\tfrac{1}{2}}h_x\right) \sinh\left(\varepsilon^{-1}\sqrt{\tfrac{1}{2}}h_y\right)}, \tag{4.41}$$

then (4.38) holds, but the boundary conditions at $x = 0$ and $y = 0$ do not hold.

Although neither of the functions satisfies the local problems exactly, they are very simple to compute as an analytic formula is given. Thus there is no need to solve local problems. Moreover, the method delivers excellent results with both choices.

If we take a particular node $I \in J_0$, and look at all elements connected to this node, then Eq. (4.41) can be used to illustrate the nodal shape functions λ_I. We obtain for $\varepsilon/h = 2, 2 \times 10^{-1}, 2 \times 10^{-3}$, the shape functions λ_I depicted in Fig. 4.2. Note that as ε approaches zero, the usual pyramid is squeezed in its domain of influence in the neighborhood around the node I. We also remark that what really matter is the relation ε/h, so if this ratio remains constant, the shape of the function is constant, but of course the local domain changes with h.

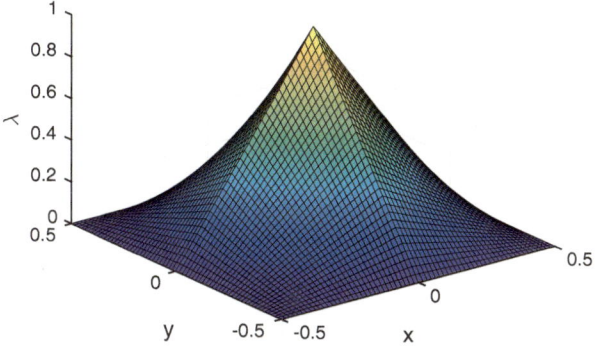

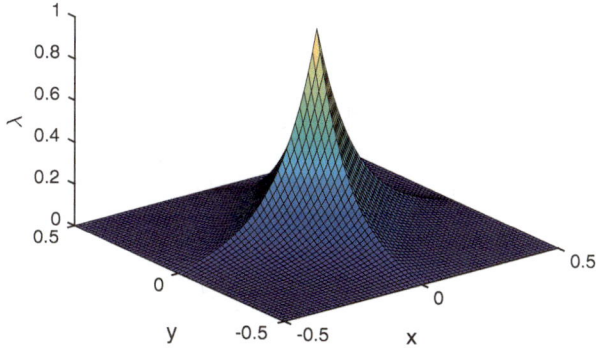

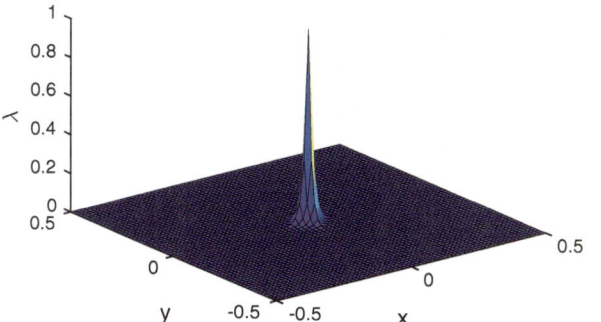

Fig. 4.2 The function λ defined by (4.41), for $\varepsilon = 1$, 10^{-1}, 10^{-3}, and $h = 1/2$

Remark 4.5 Along the lines just presented, the approach based on enriching the trial and test spaces with different functions, under the name *PGEM*, was extended to different problems [14, 104].

4.3.2.3   A Numerical Test: Source Problem

Consider the unit source problem ($f = 1$) defined on the unit square, subject to a homogeneous Dirichlet boundary condition. For a small ε, boundary layers appear close to the domain boundary. Figure 4.3 shows, for $\varepsilon^2 = 10^{-6}$, the solutions of three different methods, Galerkin, Residual Free Bubble, and the present enriched method. It is clear that the current method performs better than the other two methods.

4.4   Conclusions

In this chapter we extend asymptotic expansion techniques developed in Chap. 2 to consider two-dimensional problems. The main difference is on how to treat boundary layers, and to do so, we introduce boundary-fitted coordinates. Such system of coordinates is commonly used in problems involving geometry [72], and allows a remarkably explicit expression of the Laplacian operator. If the domain is not smooth however, such construction does not work, end expansions become much more complicate [147]. It is still possible to obtain some not so explicit results regarding the behavior of the solution, and we include an estimate for non-smooth domains.

Finally, we describe a numerical scheme that is able to approximate the exact solution without spurious oscillations. Jumping from problems in one to higher dimensions is not trivial, since element boundaries are more complicate—in one dimension the element boundary is just a point. The numerical method that we present here defines a "one-dimensional version" of the original two-dimensional operator, and use that to define local boundary conditions. Other ideas are also possible, like oversampling [85, 136].

Fig. 4.3 Comparison among Galerkin, Residual Free Bubble, and the enriched methods solving a PDE defined by the operator (4.24) with zero Dirichlet boundary conditions for $f = 1$ and $\varepsilon^2 = 10^{-6}$

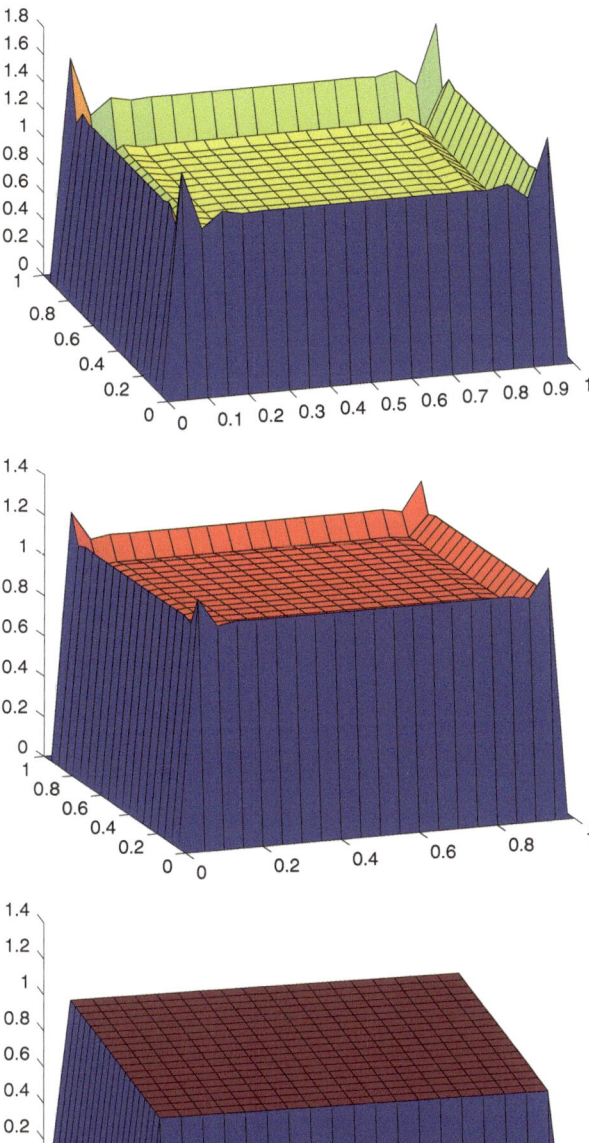

Chapter 5
Modeling PDEs in Domains with Rough Boundaries

Abstract We discuss here PDEs defined in domains where at least part of the boundary is rugous. The fully discretization of such domains can be very expensive, and we show two ways to decrease such burden. When the wrinkles are periodic, one way is to avoid the expensive discretization of the rough domain altogether, replacing the rough domain by a smooth one and changing the boundary conditions in such a way that the geometry of the wrinkles is captured. For a general domain, a possibility is to use a domain decomposition approach, solving local problems in parallel in the spirit of the Multiscale Finite Element Method. Asymptotic expansions play a key role in both alternatives, motivating the development of models, and helping in deriving error estimates.

5.1   Introduction

This chapter deals with a problem that differs from the previous ones in the sense that the small parameter is part of the domain definition. The same occurs in plate and shell problems [50, 52, 53, 63, 64, 149] or domains with small cavities [49, 56, 57, 145, 172]. The situation now is that the boundary is rough, in the sense that it contains wrinkles, which makes the analysis of the continuous problem and numerical computations challenging. An alternative to reduce the computational costs is to *homogenize* somehow the domain and make it smooth, replacing the original boundary condition with a *wall-law* that incorporates geometric aspects of the original boundary. These techniques are usually based on asymptotic expansion, and there is a vast literature on the subject [1–6, 9–12, 29, 30, 35, 159, 160].

We also consider a different approach based on a numerical scheme to deal directly with the wrinkles; see [156, 167, 171, 175, 181]. In our case we consider a finite element method of multiscale type.

Consider a domain where at least part of its boundary is rough. For instance, that is the case of a golf ball, or, to simplify the matter, domains as depicted in Fig. 5.1. It is clear that a PDE posed in such domain will depend on the geometry of the rough boundary, and any reasonably accurate discretization of the domain has to be fine enough to incorporate all the geometric information of the boundary, with high computational costs. So, a natural question is whether it is possible to avoid the over-discretization of the domain.

© The Author(s) 2017   67
A.L. Madureira, *Numerical Methods and Analysis of Multiscale Problems*,
SpringerBriefs in Mathematics, DOI 10.1007/978-3-319-50866-5_5

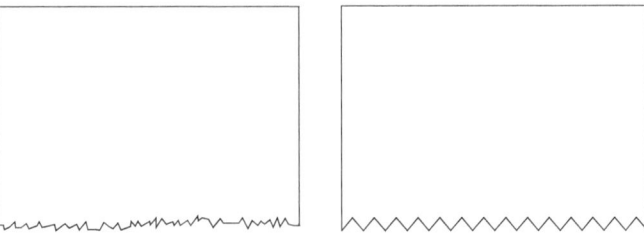

Fig. 5.1 Rough domains, with general and periodic boundaries

Fig. 5.2 The domain Ω^ϵ and its decomposition as defined by (5.1)–(5.3)

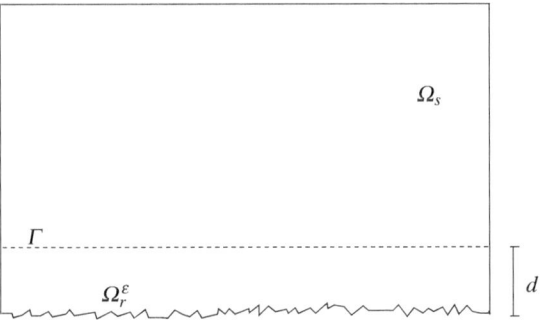

We start by simplifying the problem, and considering a domain Ω^ϵ that is, basically, a square with sides of length one, where the bottom is rugged, as in Fig. 5.1. Formally, let $\epsilon > 0$ be the characteristic length of the wrinkles, and for $d > \epsilon$, let

$$\Omega^\epsilon = \{\mathbf{x} = (x, y) \in \mathbb{R}^2 : 0 < x < 1, -d + \gamma_r^\epsilon(x) < y < 1\}, \tag{5.1}$$

where the Lipschitz function γ_r^ϵ is such that $\gamma_r^\epsilon(0) = \gamma_r^\epsilon(1) = 0$, and $\|\gamma_r^\epsilon\|_{L^\infty(0,1)} < \epsilon$. The graph of γ_r^ϵ defines Γ_r^ϵ, the bottom piece of Ω^ϵ:

$$\Gamma_r^\epsilon = \{(x, y) \in \partial\Omega^\epsilon : y = -d + \gamma_r^\epsilon(x)\}.$$

Consider two subdomains of Ω^ϵ, the first containing the wrinkles, and the second smooth; see Fig. 5.2. Let

$$\Omega_s = (0, 1) \times (0, 1), \qquad \Omega_r^\epsilon = \Omega^\epsilon \backslash \overline{\Omega_s}, \qquad \Gamma = (0, 1) \times \{0\}. \tag{5.2}$$

Note that

$$\Omega_r^\epsilon = \{\mathbf{x} = (x, y) \in \Omega^\epsilon : 0 < x < 1, -d + \gamma_r^\epsilon(x) < y < 0\}. \tag{5.3}$$

We now consider the problem

$$\begin{aligned} -\Delta u^\epsilon &= f \quad \text{in } \Omega^\epsilon, \\ u^\epsilon &= 0 \quad \text{on } \partial\Omega^\epsilon, \end{aligned} \tag{5.4}$$

for $f \in L^2(\Omega^\epsilon)$. We assume, for simplicity, that f vanishes in Ω_r^ϵ.

To tackle such problem, we start by developing an asymptotic expansion of the solution in Sect. 5.2. Then, in a first method, we develop a wall-law, homogenizing the rough boundary, replacing it by a flat one, changing the boundary conditions accordingly, in Sect. 5.3. The second method is to use finite element methods of multiscale type, in Sect. 5.4. We finally conclude with some remarks.

In Sects. 5.2 and 5.3, we assume that the rough boundary is periodic, by assuming that γ_r^ϵ is periodic. In Sect. 5.4 this is necessary only for the error analysis.

5.2 Asymptotic Expansion

The sequence u^ϵ, parametrized by $\epsilon < 1$, depends in a nontrivial way on the small parameter ϵ, and we now develop an asymptotic expansion for each solution. We start by assuming that the wrinkles are *periodic*, i.e., $\gamma_r^\epsilon(x) = \epsilon\gamma_r(\epsilon^{-1}x_1)$, where $\gamma_r : \mathbb{R} \to \mathbb{R}$ is ϵ-independent, Lipschitz continuous, and periodic with period 1. Without loss of generality, we assume that $\|\gamma_r\|_{L^\infty(\mathbb{R})} = 1$. We also assume that

$$d = d_0\epsilon, \tag{5.5}$$

where $d_0 > 1$.

5.2.1 Asymptotic Expansion Definition

The goal here is make the dependence of u^ϵ with respect to ϵ explicit. That allows not only a better understanding of the problem, but also the development of models for (5.4). We seek a formal expansion in the form

$$u^\epsilon \sim u^0 + \epsilon v^1(\epsilon) + \epsilon\Psi^1(\epsilon) + \epsilon^2 v^2(\epsilon) + \epsilon^2\Psi^2(\epsilon) + \cdots, \tag{5.6}$$

where $\Psi^i(\epsilon)$ are boundary correctors. The terms $v^i(\epsilon)$ and $\Psi^i(\epsilon)$ depend on ϵ, as the notation indicates. It might appear strange at a first sight, but its introduction makes the development of the expansion simpler. At a second stage, it is possible to reorder (5.6) such that the terms are ϵ-independent.

The procedure to find the terms of the expansion involves a domain decomposition approach, and it is based on the following result.

Fig. 5.3 Domain
decomposition

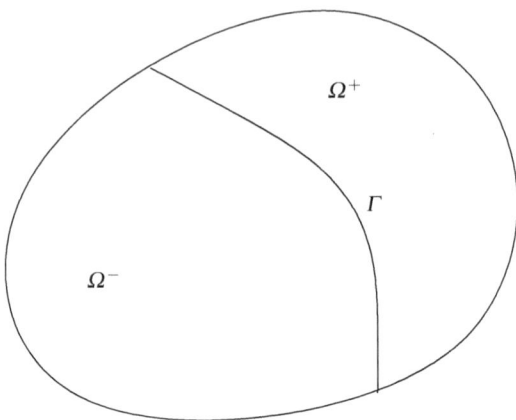

Lemma 5.1 *Let $\Omega \subset \mathbb{R}^2$ be a bounded Lipschitz domain, and consider the interface Γ dividing Ω into two subdomains Ω^- and Ω^+ (Fig. 5.3). Let the function e be such that $e = 0$ in $(\partial\Omega^- \backslash \Gamma) \cup (\partial\Omega^+ \backslash \Gamma)$. Then, there exists a constant c, that does not depend on Ω^- such that*

$$\|e\|_{H^1(\Omega^-)} + \|e\|_{H^1(\Omega^+)} \le c \Big(\|\Delta e\|_{L^2(\Omega^-)}$$

$$+ \|\Delta e\|_{L^2(\Omega^+)} + \|J(e)\|_{H^{\frac{1}{2}}(\Gamma)} + \|J(\partial e/\partial n)\|_{H^{-\frac{1}{2}}(\Gamma)} \Big). \qquad (5.7)$$

Here, $J(\cdot)$ is the jump function with respect to Γ.

Proof Let $e^- = e|_{\Omega^-}$, $e^+ = e|_{\Omega^+}$. It follows from Green's identity that

$$\int_{\Omega^-} |\nabla e^-|^2 \, d\mathbf{x} = - \int_{\Omega^-} e^- \Delta e^- \, d\mathbf{x} - \int_{\Gamma} e^- \frac{\partial e^-}{\partial n^+} \, d\Gamma,$$

$$\int_{\Omega^+} |\nabla e^+|^2 \, d\mathbf{x} = - \int_{\Omega^+} e^+ \Delta e^+ \, d\mathbf{x} + \int_{\Gamma} e^+ \frac{\partial e^+}{\partial n^+} \, d\Gamma,$$

where n^+ is the outward normal with respect to $\partial\Omega^+$. Combining the identities, and adding and subtracting $\int_{\Gamma} e^- \partial e^+/\partial n^+ \, d\Gamma$, we gather that

$$|e^-|^2_{H^1(\Omega^-)} + |e^+|^2_{H^1(\Omega^+)}$$

$$= - \int_{\Omega^-} e^- \Delta e^- \, d\mathbf{x} - \int_{\Omega^-} e^+ \Delta e^+ \, d\mathbf{x} + \int_{\Gamma} e^- \left(\frac{\partial e^+}{\partial n^+} - \frac{\partial e^-}{\partial n^+} \right) d\Gamma$$

$$+ \int_{\Gamma} (e^+ - e^-) \frac{\partial e^+}{\partial n^+} \, d\Gamma$$

$$\leq \|e^-\|_{L^2(\Omega-)} \|\Delta e^-\|_{L^2(\Omega-)} + \|e^+\|_{L^2(\Omega+)} \|\Delta e^+\|_{L^2(\Omega+)}$$

$$+ \|e^-\|_{H^{\frac{1}{2}}(\Gamma)} \left\| J\left(\frac{\partial e}{\partial n}\right) \right\|_{H^{-\frac{1}{2}}(\Gamma)} + \|J(e)\|_{H^{\frac{1}{2}}(\Gamma)} \left\| \frac{\partial e}{\partial n}^+ \right\|_{H^{-\frac{1}{2}}(\Gamma)}$$

$$\leq (\|\Delta e^-\|_{L^2(\Omega-)} + \|\Delta e^+\|_{L^2(\Omega+)}) \|e\|_{L^2(\Omega)}$$

$$+ \|e^-\|_{H^1(\Omega-)} \left\| J\left(\frac{\partial e}{\partial n}\right) \right\|_{H^{-\frac{1}{2}}(\Gamma)} + \|J(e)\|_{H^{\frac{1}{2}}(\Gamma)} \|e^+\|_{H^1(\Omega+)}.$$

We used above the Cauchy–Schwartz inequality, the trace inequality, and the duality between $H^{-1/2}(\Gamma)$ and $H^{1/2}(\Gamma)$. The lemma then follows from a direct application of the Poincaré's inequality. □

We next apply Lemma 5.1 along with the decomposition described in Sect. 5.1. It is only natural to define u^0, the first term of the expansion, by

$$\begin{aligned} -\Delta u^0 &= f \quad \text{in } \Omega_s, \\ u^0 &= 0 \quad \text{on } \partial\Omega_s, \qquad u^0 = 0 \quad \text{on } \Omega_r^\epsilon. \end{aligned} \tag{5.8}$$

From Lemma 5.1 with $e = u^\epsilon - u^0$, we gather that the error originates from the jump $J(\partial_{x_2} u^0)$:

$$\|e\|_{H^1(\Omega_r^\epsilon)} + \|e\|_{H^1(\Omega_s)} \leq c \|\partial_{x_2} u^0\|_{H^{-\frac{1}{2}}(\Gamma)}.$$

To fix this, consider $\phi^0(x_1) = \partial_{x_2} u^0(x_1, 0)$, i.e., ϕ^0 is the restriction of $\partial u^0/\partial x_2|_{\Omega_s}$ on Γ. Let χ_r^ϵ be the characteristic function of the set Ω_r^ϵ, i.e.,

$$\chi_r^\epsilon(\mathbf{x}) = \begin{cases} 1 & \text{if } \mathbf{x} \in \Omega_r^\epsilon, \\ 0 & \text{otherwise.} \end{cases}$$

Adding $-\chi_r^\epsilon(\mathbf{x})(x_2 + d)\phi^0(x_1)$ to the expansion, we correct the error due to the jump of the derivative in the x_2 direction. Since the Dirichlet condition at Γ_r^ϵ is not satisfied, we add the corrector $\Psi^1(\epsilon) - \chi_r^\epsilon Z^1(\epsilon)$ to the expansion, where

$$\begin{aligned} -\Delta\Psi^1(\epsilon) &= -\chi_r^\epsilon \Delta[(\epsilon^{-1}x_2 + d_0)\phi^0 + Z^1(\epsilon)] \quad \text{in } \Omega^\epsilon, \\ \Psi^1(\epsilon) &= (\epsilon^{-1}x_2 + d_0)\phi^0 + Z^1(\epsilon) \quad \text{on } \Gamma_r^\epsilon. \end{aligned} \tag{5.9}$$

The extra term $Z^1(\epsilon)$ is important to guarantee the exponential decay of $\Psi^1(\epsilon)$ to zero, in the x_2 direction. Note that $\Psi^1(\epsilon)$ and $Z^1(\epsilon)$ depend on ϵ, and actually they are not even well defined yet. In general, we define $\Psi^i(\epsilon)$ and $Z^i(\epsilon)$ as formal expansions with respect to ϵ,

$$\Psi^i(\varepsilon)(\mathbf{x}) \sim \psi^0(\varepsilon^{-1}\mathbf{x})\phi^{i-1}(x_1) + \varepsilon\psi^1(\varepsilon^{-1}\mathbf{x})\frac{\partial\phi^{i-1}}{\partial x_1}(x_1)$$

$$+ \varepsilon^2\psi^2(\varepsilon^{-1}\mathbf{x})\frac{\partial^2\phi^{i-1}}{\partial x_1^2}(x_1) + \cdots,$$

$$Z^i(\varepsilon)(\mathbf{x}) \sim z^0\phi^{i-1}(x_1) + \varepsilon z^1\frac{\partial\phi^{i-1}}{\partial x_1}(x_1) + \varepsilon^2 z^2\frac{\partial^2\phi^{i-1}}{\partial x_1^2}(x_1) + \cdots.$$

Above, ψ^i is periodic with period ε^{-1} in the x_1 direction, and z^i depend only on the geometry of the wrinkle, and, in particular, are ε-independent. We postpone the exact definition of these terms to Sect. 5.2.2.

Now $e = u^\varepsilon - \left[u^0 - \chi_r^\varepsilon(x_2 + d)\phi^0 - \varepsilon\chi_r^\varepsilon Z^1(\varepsilon) + \varepsilon\Psi^1(\varepsilon)\right]$, and we have that

$$\|e\|_{H^1(\Omega_r^\varepsilon)} + \|e\|_{H^1(\Omega_s)} \leq \varepsilon\|Z^1(\varepsilon) + d_0\phi^0\|_{H^{\frac{1}{2}}(\Gamma)}.$$

We continue to define the terms in the expansion. Let

$$\begin{aligned}
-\Delta u^1 &= 0 \quad \text{in } \Omega_s, & u^1 &= -(d_0 + z^0)\phi^0 \quad \text{on } \Gamma, \\
u^1 &= 0 \quad \text{on } \partial\Omega_s\backslash\Gamma, & u^1 &= 0 \quad \text{on } \Omega_r^\varepsilon.
\end{aligned} \tag{5.10}$$

Hence, if $e = u^\varepsilon - \left[u^0 - (x_2 + d)\chi_r^\varepsilon\phi^0 - \varepsilon Z^1(\varepsilon)\chi_r^\varepsilon + \varepsilon\Psi^1(\varepsilon) + \varepsilon u^1\right]$, then

$$\|e\|_{H^1(\Omega_r^\varepsilon)} + \|e\|_{H^1(\Omega_s)} \leq \varepsilon\left\|\frac{\partial u^1}{\partial x_2}\right\|_{H^{-\frac{1}{2}}(\Gamma)} + \varepsilon^2\left\|z^1\frac{\partial\phi^0}{\partial x_1} + \varepsilon^2\frac{\partial^2\phi^0}{\partial x_1^2} + \cdots\right\|_{H^{\frac{1}{2}}(\Gamma)}.$$

Let $\phi^1 = \partial u^1/\partial x_2|_\Gamma$, and incorporate $-\varepsilon\chi_r^\varepsilon(x_2 + d)\phi^1 + \varepsilon^2\Psi^2(\varepsilon) - \varepsilon^2\chi_r^\varepsilon Z^2(\varepsilon)$ to the expansion, where

$$\begin{aligned}
-\Delta\Psi^2(\varepsilon) &= -\chi_r^\varepsilon\Delta[(\varepsilon^{-1}x_2 + d_0)\phi^1 + Z^2(\varepsilon)] \quad \text{in } \Omega^\varepsilon, \\
\Psi^2(\varepsilon) &= (\varepsilon^{-1}x_2 + d_0)\phi^1 + Z^2(\varepsilon) \quad \text{on } \Gamma_r^\varepsilon.
\end{aligned} \tag{5.11}$$

We now have

$$\begin{aligned}
e = u^\varepsilon - \big[&u^0 - (x_2 + d)\chi_r^\varepsilon\phi^0 - \varepsilon Z^1(\varepsilon)\chi_r^\varepsilon + \varepsilon\Psi^1(\varepsilon) + \varepsilon u^1 \\
&- \varepsilon(x_2 + d)\chi_r^\varepsilon\phi^1 + \varepsilon^2\Psi^2(\varepsilon) - \varepsilon^2\chi_r^\varepsilon Z^2(\varepsilon)\big],
\end{aligned}$$

and

$$\|e\|_{H^1(\Omega_r^\varepsilon)} + \|e\|_{H^1(\Omega_s)} \leq \varepsilon^2\left\|Z^2(\varepsilon) + d_0\phi^1 + z^1\frac{\partial\phi^0}{\partial x_1} + \varepsilon z^2\frac{\partial^2\phi^0}{\partial x_1^2} + \cdots\right\|_{H^{\frac{1}{2}}(\Gamma)}.$$

Proceeding yet a bit further by defining $\phi^2 = \partial u^2/\partial x_2|_\Gamma$, we add

$$\varepsilon^2 u^2 - \varepsilon^2(x_2 + d)\chi_r^\varepsilon\partial_{x_2}u^2 + \varepsilon^3\Psi^3(\varepsilon) - \varepsilon^3\chi_r^\varepsilon Z^3(\varepsilon)$$

to the expansion, where

$$-\Delta u^2 = 0 \quad \text{in } \Omega_s, \qquad u^2 = 0 \quad \text{on } \Omega_r^\epsilon,$$
$$u^2 = -(d_0 + z^0)\phi^1 + z^1 \frac{\partial \phi^0}{\partial x_1} \quad \text{on } \Gamma, \qquad u^2 = 0 \quad \text{on } \partial\Omega_s \backslash \Gamma, \tag{5.12}$$

and

$$-\Delta \Psi^3(\varepsilon) = -\chi_r^\epsilon \Delta[(\varepsilon^{-1}x_2 + d_0)\phi^2 + Z^3(\varepsilon)] \quad \text{in } \Omega^\epsilon,$$
$$\Psi^3(\varepsilon) = (\varepsilon^{-1}x_2 + d_0)\phi^2 + Z^3(\varepsilon) \quad \text{on } \Gamma_r^\epsilon. \tag{5.13}$$

The error is now

$$e = u^\epsilon - \big[u^0 - (x_2 + d)\chi_r^\epsilon \phi^0 + \varepsilon \Psi^1(\varepsilon) - \varepsilon \chi_r^\epsilon Z^1(\varepsilon) + \varepsilon u^1$$
$$- \varepsilon(x_2 + d)\chi_r^\epsilon \phi^1 + \varepsilon^2 \Psi^2(\varepsilon) - \varepsilon^2 \chi_r^\epsilon Z^2(\varepsilon)$$
$$+ \varepsilon^2 u^2 - \varepsilon^2 (x_2 + d)\chi_r^\epsilon \partial_{x_2} u^2 + \varepsilon^3 \Psi^3(\varepsilon) - \varepsilon^3 \chi_r^\epsilon Z^3(\varepsilon)\big],$$

and

$$\|e\|_{H^1(\Omega_f^\epsilon)} + \|e\|_{H^1(\Omega_s)} \leq \varepsilon^3 \left\| z^2 \frac{\partial^2 \phi^0}{\partial x_1^2}\phi^0 + z^1 \frac{\partial \phi^1}{\partial x_1} + z^0 \phi^2 + \varepsilon \ldots \right\|_{H^{\frac{1}{2}}(\Gamma)}.$$

The additional terms of the expansions are added in similar form.

5.2.2 The Boundary Corrector Problem

We investigate now boundary corrector problems in further details. Consider the problem of finding functions $\varXi$ and Z such that

$$-\Delta \varXi = -\chi_r^\epsilon \Delta[(\varepsilon^{-1}x_2 + d_0)\phi + Z) \quad \text{in } \Omega^\epsilon,$$
$$\varXi = (\varepsilon^{-1}x_2 + d_0)\phi + Z \quad \text{on } \Gamma_r^\epsilon. \tag{5.14}$$

Here, while ϕ is a given function depending on x_1 only, Z is unknown, and it is introduced to guarantee that $\varXi$ decays exponentially to zero. Note that (5.9), (5.11), and (5.13) are of the form (5.14), for $\varXi = \Psi^i$ and $\phi = \phi^1$.

Both $\varXi$ and Z depend on ε, and we assume that

$$\varXi \sim \varXi^0 + \varepsilon \varXi^1 + \varepsilon^2 \varXi^2 + \cdots, \qquad Z \sim Z^0 + \varepsilon Z^1 + \varepsilon^2 Z^2 + \cdots. \tag{5.15}$$

The functions in (5.14) are not periodic in general. However, we exploit the periodicity of the wrinkles and rewrite the corrector problem as a sequence of problems posed in periodic domains. Moreover, motivated by the oscillatory behavior of the wrinkles, we use the scaled variables

$$\hat{\mathbf{x}} = (\hat{x}_1, \hat{x}_2) = (\varepsilon^{-1} x_1, \varepsilon^{-1} x_2),$$

and assume that a first approximation for $\varXi$ and Z is given by

$$\varXi^0(\mathbf{x}) = \xi^0(\hat{x}_1, d_0 + \hat{x}_2)\phi(x_1), \quad Z^0(x_1) = z^0\phi(x_1), \tag{5.16}$$

where ξ^0 is $\hat{x}_1$-periodic, z^0 is constant, and $\varXi^0$ and z^0 are to be determined.

Remark 5.1 At this stage, there are no good arguments indicating that z^0 is constant. Note, however, that the simpler Z^0 is, the better. *A posteriori* we show that the assumption on Z^0 is enough for the exponential decay of $\varXi^0$.

A straightforward computation shows that the Laplacian of a function in the form $u(\hat{x}_1, \hat{x}_2)v(x_1)$ is given by

$$-\Delta_{\mathbf{x}}(uv) = -\varepsilon^{-2}(\Delta_{\hat{\mathbf{x}}} u)v - 2\varepsilon^{-1}\frac{\partial u}{\partial \hat{x}_1}v' - uv''. \tag{5.17}$$

Thus,

$$\begin{aligned}
-\Delta(\xi^0\phi) &= -\varepsilon^{-2}\big(\Delta_{\hat{\mathbf{x}}}\xi^0\big)\phi - 2\varepsilon^{-1}\frac{\partial \xi^0}{\partial \hat{x}_1}\phi' - \xi^0\phi'', \\
-\Delta(\hat{x}_2\phi) &= -\hat{x}_2\phi'', \qquad -\Delta(z^0\phi) = -z^0\phi''.
\end{aligned} \tag{5.18}$$

To continue the description of the expansion, we introduce the *cell problem*, typical of homogenization theory. Usually these cell problems carry "information" from the microscale to the macroscopic behavior of the solution.

In the present case, the cell problem is defined in the semi-infinite domain Ω_r, defined by the wrinkle geometry. Indeed, Ω_r is the strip bounded by $\hat{x}_1 = 0$ and $\hat{x}_1 = 1$, with lower bound given by $\Gamma_r^- = \{(\hat{x}_1, \gamma_r(\hat{x}_1)) : \hat{x}_1 \in (0, 1)\}$, i.e.,

$$\Omega_r = \{(\hat{x}, \hat{y}) \in \mathbb{R}^2 : \hat{x} \in (0, 1), \ \hat{y} > \gamma_r(\hat{x})\}.$$

See Fig. 5.4.

We define $C_{\mathrm{per}}^\infty(\Omega_r)$ by restricting to Ω_r the functions in $C^\infty(\mathbb{R}^2)$ that are 1-periodic with respect to $\hat{\theta}$. Let $H_{\mathrm{per}}^1(\Omega_r)$ be the closure of $C_{\mathrm{per}}^\infty(\Omega_r)$ with respect to the $H^1(\Omega_r)$ norm. We also introduce the space of exponentially decaying functions

$$S(\Omega_r) = \{w \in H_{\mathrm{per}}^1(\Omega_r) : w(\hat{x}, \hat{y})e^{\alpha\hat{y}} \in H^1(\Omega_r) \text{ for some } \alpha > 0\}.$$

Based in (5.14), (5.15), (5.16), and (5.18), and formally comparing similar powers of ε, we conclude that ξ^0 is harmonic. The boundary conditions over the

Fig. 5.4 The cell

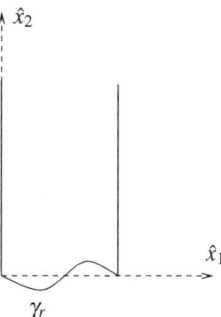

wrinkles come from (5.14), (5.15), and (5.16). It is possible to show [9, 160] that there exists a unique $\xi^0 \in S(\Omega_r)$, and a unique constant z^0 such that

$$\partial_{\hat{x}_1\hat{x}_1}\xi^0 + \partial_{\hat{x}_2\hat{x}_2}\xi^0 = 0 \quad \text{in } \Omega_r, \qquad \xi^0 = \hat{x}_2 + z^0 \quad \text{on } \Gamma_r^-,$$
$$\xi^0 \text{ is } \hat{x}_1 - \text{periodic}, \qquad \lim_{\hat{x}_2 \to \infty} \xi^0 = 0. \tag{5.19}$$

Moreover, $0 \le z \le \|\gamma_r\|_{L^\infty(0,1)}$, and $\|w\|_{H^1(\Omega_r)} \le c$, where c depends only on the geometry of the wrinkles.

We have then that

$$-\Delta[(\xi^0 - \chi_r^\epsilon\hat{x}_2 + \chi_r^\epsilon z^0)\phi] = -2\varepsilon^{-1}\partial_{\hat{x}_1}\xi^0\phi' - \xi^0\phi'' + \chi_r^\epsilon\hat{x}_2\phi'' - \chi_r^\epsilon z^0\phi'' \quad \text{in } \Omega^\epsilon,$$
$$\xi^0 - \hat{x}_2 + z^0 = 0 \quad \text{on } \Gamma_r^\epsilon.$$

Let $\Xi^1(\mathbf{x}) = \xi^1(\hat{x}_1, d_0 + \hat{x}_2)\phi'(x_1)$, $Z^1 = z^1\phi'$, where

$$-[\partial_{\hat{x}_2\hat{x}_2}\xi^1 + \partial_{\hat{x}_1\hat{x}_1}\xi^1] = 2\partial_{\hat{x}_1}\xi^0 \quad \text{in } \Omega_r, \qquad \xi^1 = z^1 \quad \text{on } \Gamma_r^-.$$

Then,

$$-\Delta[(\xi^0 - \chi_r^\epsilon\hat{x}_2 + \chi_r^\epsilon z^0)\phi + \varepsilon(\xi^1 + \chi_r^\epsilon z^1)\phi']$$
$$= -\xi^0\phi'' + \chi_r^\epsilon\hat{x}_2\phi'' - \chi_r^\epsilon z^0\phi'' - 2\partial_{\hat{x}_1}\xi^1\phi' - \varepsilon\xi^1\phi''' - \varepsilon\chi_r^\epsilon z^1\phi''' \quad \text{in } \Omega^\epsilon,$$
$$(\xi^0 - \chi_r^\epsilon\hat{x}_2 + \chi_r^\epsilon z^0)\phi + \varepsilon(\xi^1 + \chi_r^\epsilon z^1)\phi' = 0 \quad \text{on } \Gamma_r^\epsilon.$$

Analogously, $\Xi^2(\mathbf{x}) = \xi^2(\hat{x}_1, d_0 + \hat{x}_2)\phi''(x_1)$, and $Z^2 = z^2\phi''$ where

$$-[\partial_{\hat{x}_2\hat{x}_2}\xi^2 + \partial_{\hat{x}_1\hat{x}_1}\xi^2] = \xi^0 - \chi_r\hat{x}_2 + \chi_r z^0 + 2\partial_{\hat{x}_1}\xi^1 \quad \text{in } \Omega_r, \qquad \xi^2 = z^2 \quad \text{on } \Gamma_r^-,$$

where $\chi_r(\hat{x}_2) = 1$ if $\hat{x}_2 \le 0$, and $\chi_r(\hat{x}_2) = 0$ otherwise. It is easy to see the right-hand side of the equations become more involved as we proceed with the definition of the expansion terms. Note, however, that the equations do not involve ϕ or its derivatives.

We finally conclude that

$$\Xi(\varepsilon) \sim \xi^0 \phi + \varepsilon \xi^1 \phi' + \varepsilon^2 \xi^2 \phi'' + \cdots, \qquad Z(\varepsilon) \sim z^0 \phi + \varepsilon z^1 \phi' + \varepsilon^2 z^2 \phi'' + \cdots.$$
$$(5.20)$$

5.3 Derivation of Wall-Laws

Our goal is to approximate u^ε using finite elements (or finite differences), without having to discretize the rough boundary. One first step would be to try to approximate u^ε in Ω_s only, since this is a smooth domain. Heuristically, that means that we consider only the functions that actually have influence in the interior of the domain, i.e.,

$$u \approx u^0 + \varepsilon u^1.$$

The idea is to use the above approximation, but without solving all the problems that define each term. The functions u^0, u^1 are defined by (5.8), (5.10), and note that, over Γ, $u \approx u^0 + \varepsilon u^1 = -\varepsilon(d_0 + z^0)\phi^0$, and $\partial_{x_2} u \approx \phi^0 + \varepsilon \phi^1$. So

$$u + \varepsilon(d_0 + z^0)\partial_{x_2} u \approx \varepsilon^2 (d_0 + z^0)\phi^1,$$

and this amount can be small enough for certain applications.

Based on the above arguments, define $\bar{u}$ approximating u^ε in Ω_s by

$$-\Delta\bar{u} = f \quad \text{in } \Omega_s,$$
$$\bar{u} + \varepsilon(d_0 + z^0)\partial_{x_2}\bar{u} = 0 \quad \text{on } \Gamma, \qquad \bar{u} = 0 \quad \text{on } \partial\Omega_s \backslash \Gamma.$$
$$(5.21)$$

The error estimates follow by developing asymptotic expansions for $\bar{u}$. In fact, it is easy to see that

$$\bar{u} \sim \bar{u}^0 + \varepsilon \bar{u}^1 + \varepsilon^2 \bar{u}^2 + \cdots,$$

where

$$-\Delta\bar{u}^i = \delta_{i,0} f \quad \text{in } \Omega_s,$$
$$\bar{u}^i = -(c_0 + z^0)\partial_{x_2}\bar{u}^{i-1} \quad \text{on } \Gamma, \qquad \bar{u}^i = 0 \quad \text{on } \partial\Omega_s \backslash \Gamma.$$
$$(5.22)$$

Also,

$$\left\| \bar{u} - \sum_{i=0}^{n} \varepsilon^i \bar{u}^i \right\|_{H^k(\Omega_s)} \leq c\varepsilon^{n+1}.$$
$$(5.23)$$

The modeling error estimates are then as follows

$$\|u^\epsilon - \bar{u}\|_{H^1(\Omega_s)} \le \|u^\epsilon - u^0 - \epsilon u^1\|_{H^1(\Omega_s)} + \|\bar{u} - \bar{u}^0 - \epsilon\bar{u}^1\|_{H^1(\Omega_s)}$$

$$\le \|u^\epsilon - u^0 - \epsilon u^1 - \epsilon\psi^0\phi^0\|_{H^1(\Omega_s)} + \epsilon\|\psi^0\phi^0\|_{H^1(\Omega_s)} + c\epsilon^2 \le c\epsilon^{1/2},$$

where in the first inequality we used the triangle inequality, and the identities $u^0 = \bar{u}^0$ and $u^1 = \bar{u}^1$.

In a similar fashion, we can consider other norms, for instance

$$\|u^\epsilon - \bar{u}\|_{L^2(\Omega_s)} \le c\epsilon^2,$$

Another important measure is how well our model approximates the exact solution in the interior of the domain, i.e., consider $\mathring{\Omega}_s \subset \Omega_s$ such that $\overline{\mathring{\Omega}}_s \cap \overline{\Gamma} = \emptyset$. Hence

$$\|u^\epsilon - \bar{u}\|_{L^2(\Omega_s)} + \|u^\epsilon - \bar{u}\|_{H^k(\mathring{\Omega}_s)} \le c\epsilon^2.$$

This improved convergence is due to the fact that the boundary layer has no influence far from the wrinkles.

We also compare in Tables 5.1 and 5.2 the approximation properties of our model and a simple minded model, which approximates u^ϵ by u^0 only. We call such model as "order zero model," and our model as "order one model." Note the improved convergence rate for most of the norms, with the exception of the $H^1(\Omega_s)$ norm. This is due to the fact that neither approaches captures the boundary layer exactly.

Remark 5.2 Assuming periodic wrinkles, several authors proposed wall-laws for different operators [1–6, 10–12, 29, 35, 159, 160]. The general procedure involves cell problems, as described above. In particular, the case of wrinkled curved boundaries was considered in [159, 160]. For a random case, see [30].

Table 5.1 Relative error convergence rates for the order zero model

Quantity	$L^2(\Omega_s)$	$L^2(\mathring{\Omega}_s)$
u	$O(\epsilon)$	$O(\epsilon)$
∇u	$O(\epsilon^{1/2})$	$O(\epsilon)$

The domain $\mathring{\Omega}_s$ is such that $\overline{\mathring{\Omega}}_s \subset \Omega_s$, i.e., the rates with respect to the $L^2(\mathring{\Omega}_s)$ are interior error estimates

Table 5.2 Relative error convergence rates for order one model

Quantity	$L^2(\Omega_s)$	$L^2(\mathring{\Omega}_s)$
u	$O(\epsilon^2)$	$O(\epsilon^2)$
∇u	$O(\epsilon^{1/2})$	$O(\epsilon^2)$

Again, the rates with respect to $L^2(\mathring{\Omega}_s)$ are interior error estimates—see captions of Table 5.1

5.4   A Multiscale Finite Element Method

Here we propose and analyze a finite element scheme of multiscale type to deal
with PDEs posed in domains with rough boundaries. The method is quite general
since there is no need to assume any periodicity of the boundaries. However, if the
boundaries are periodic, we can prove convergence of the method.

Method such as the one described here fills a gap in terms of cost effective
approximations for rough boundary problems, since wall-laws are derived based on
restrictive assumptions on the boundary, as periodicity or certain random properties.

On the other hand, as discussed in Chap. 6, there are several numerical methods
for PDEs with oscillatory coefficients [24, 73, 74, 78, 85, 87, 134, 136–138, 188].
A common feature among these methods is that they can be defined independently
of the periodicity assumptions (albeit some tune up can be made in some special
cases).

The goal here is to discuss a finite element method of multiscale type for
problems posed in domains with rough boundaries, as (5.4).

5.4.1   Method Definition

The simplest instance of multiscale finite element method is based on a Galerkin
approach, but with modified basis functions [85, 87, 134, 136–138]. Such basis
functions are usually local, elementwise, solutions of the original problem, with
the goal of upscaling the small scale information.

We propose here a method where the basis functions have local support, as in the
traditional finite element method. However, we forsake the polynomial spaces, and,
in general, not even the elements are polygons since, close to the boundary, they can
have a rough geometry. The basic idea is that the influence of the geometry should
be capture by the basis functions without increasing the size of the final system.
Thus, the scheme is a Galerkin method where the basis functions are local solutions
of the PDE, possibly in a rough element.

Let $N \in \mathbb{N}$ and $h = 1/(N + 1) > \varepsilon$. For $i = 0, \ldots, N$, let $K_i^\epsilon = \{(x, y) \in \Omega_r^\epsilon : ih < x < (i + 1)h\}$. Note that these elements define a partition of Ω_r^ϵ.
Next, consider a Cartesian mesh for Ω_s employing quadrilaterals of size h. This
whole procedure introduces a finite element partition $\mathscr{T}_h$ for Ω^ϵ, not all of them
being quadrilaterals. Indeed, if $K \in \mathscr{T}_h \cap \Omega_s$, then K is a quadrilateral of size h.
Otherwise, K might have a rough side, as in Fig. 5.5. The set of mesh nodal points
of $\mathscr{T}_h$ is $\mathscr{N} = \{(ih, jh) \in \Omega^\epsilon : i = 1, \ldots, N, j = 0, \ldots, N\}$.

For each node $\mathbf{x}_i \in \mathscr{N} \cap \Omega_s$, we associate a piecewise bilinear function $\lambda_i \in H_0^1(\Omega^\epsilon)$, such that $\lambda_i(\mathbf{x}_j) = \delta_{ij}$ for all $\mathbf{x}_j \in \mathscr{N}$. For the nodes $\mathbf{x}_i \in \mathscr{N} \cap \Gamma$, define
$\lambda_i \in H_0^1(\Omega^\epsilon)$ such that

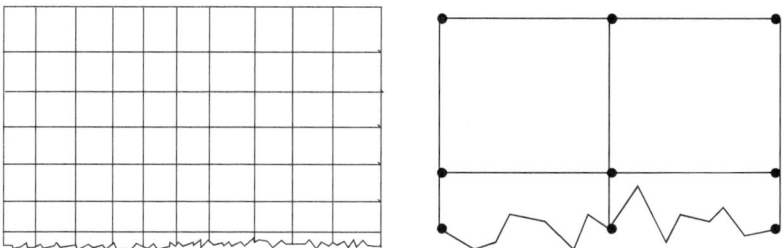

Fig. 5.5 A mesh for Ω^ϵ, and a patch of elements of $\mathscr{T}_h$ intercepting Γ_r^ϵ

$$
\begin{aligned}
-\Delta \lambda_i &= 0 \quad \text{in } \cup_{j=0}^{N} K_j^\epsilon, \\
\lambda_i(\mathbf{x}_j) &= \delta_{ij} \quad \text{for all } \mathbf{x}_j \in \mathcal{N}, \qquad \lambda_i \text{ is linear on } \cup_{j=0}^{N} \partial K_j^\epsilon \cap \Omega^\epsilon.
\end{aligned} \tag{5.24}
$$

Then, extend λ_i to Ω_s by imposing that λ_i is piecewise bilinear in Ω_s.

With the above functions, define

$$
V_h^\epsilon = \operatorname{span}\{\lambda_i\} \subset H_0^1(\Omega^\epsilon).
$$

The multiscale solution $u_h^\epsilon \in V_h^\epsilon$ is simply the Galerkin approximation of u^ϵ in V_h^ϵ, i.e,

$$
\int_{\Omega^\epsilon} \nabla u_h^\epsilon(\mathbf{x}) \cdot \nabla v_h(\mathbf{x}) \, d\mathbf{x} = \int_{\Omega^\epsilon} f(\mathbf{x}) v_h(\mathbf{x}) \, d\mathbf{x} \qquad \text{for all } v_h \in V_h^\epsilon. \tag{5.25}
$$

Remark 5.3 Note that the basis functions depend on ε, and their computation might be expensive. It is, however, possible to reduce the cost of the method using parallel computation. Also, at least for two-dimensional problems, it is possible to reduce the final matrix band by numbering the nodes in an appropriate way (Sarkis, Private Communication, 2007).

Remark 5.4 The present method is particularly attractive if (5.4) has to be solved several times, for different right-hand sides. Indeed, the basis functions have to be computed only once, and the size of the final system is ε-independent.

5.4.2   Numerical Analysis

Here we develop an error analysis for our scheme. The analysis is based on asymptotic methods, see Sect. 5.2, and is restricted to the periodic case.

We assume that there exist positive constants $\gamma \in (0, 1]$ and c_0 such that

$$c_0 \varepsilon^{1-\gamma} \le d. \tag{5.26}$$

Compare with (5.5). We recall (5.1), where it is clear that d is part of the definition of Ω^ϵ. See Fig. 5.2, and also (5.3).

Remark 5.5 The restriction (5.26) is related to the fact that d should not be too small. Indeed, the rough boundary generates oscillations in u^ϵ, which decay exponentially with y/ε, and (5.26) guarantees that the oscillatory part of the solution is at most polynomial in Ω_s, with respect to ε.

5.4.2.1   Asymptotic Expansion of the Exact Solution

We slightly redefine the asymptotic expansion for u^ϵ, that is different from Sect. 5.2, due to (5.26):

$$u^\epsilon(\mathbf{x}) \sim \begin{cases} (d + y - \varepsilon W - \varepsilon z)\dfrac{\partial u^0}{\partial y}(x, 0) + (d + y)\dfrac{\partial u^1}{\partial y}(x, 0) + \cdots & \text{in } \Omega_r^\epsilon, \\[2mm] u^0(\mathbf{x}) + u^1(\mathbf{x}) + \cdots & \text{in } \Omega_s. \end{cases} \tag{5.27}$$

We define u^0 as in (5.8) and $W(x, y) = \psi^0(\varepsilon^{-1}x, \varepsilon^{-1}(d + y))$ in Ω_r^ϵ.
Finally, let

$$\begin{aligned} -\Delta u^1 &= 0 \quad \text{in } \Omega_s, & u^1 &= (d - \varepsilon z^0)\dfrac{\partial u^0}{\partial y} \quad \text{on } \Gamma, \\ u^1 &= 0 \quad \text{on } \partial\Omega_s \backslash \Gamma, & u^1 &= 0 \quad \text{on } \Omega_r^\epsilon. \end{aligned} \tag{5.28}$$

For

$$e(\mathbf{x}) = u^\epsilon(\mathbf{x}) - u^0(\mathbf{x}) - u^1(\mathbf{x}) + (y - \varepsilon W)\frac{\partial u^0}{\partial y}(x, 0)\chi_r^\epsilon,$$

under the hypothesis (5.26), there exists a constant c that is independent of ε such that

$$\|e\|_{H^1(\Omega^\epsilon)} \le cd^{3/2}, \qquad \|e\|_{H^1(\Omega_s)} \le cd^2. \tag{5.29}$$

The following estimates also hold.

Lemma 5.2 *Let u^0, u^1, and W be as above. Then there exists an ε-independent constant c such that*

$$\|u^0\|_{H^1(\Omega_s)} + d^{-1}\|u^1\|_{H^1(\Omega_s)} + \varepsilon^{-1/2}\|W\|_{L^2(\Omega_r^\epsilon)} + \varepsilon^{1/2}\|\nabla W\|_{L^2(\Omega_r^\epsilon)} \le c.$$

We now note that the Poincaré's inequality holds uniformly with respect to ε [160], i.e., there exists an ε-independent constant c such that

$$\|v\|_{L^2(\Omega^\varepsilon)} \le c|v|_{H^1(\Omega^\varepsilon)},$$

for all $v \in H^1_0(\Omega^\varepsilon)$. We conclude then that Céa's Lemma holds uniformly with respect to ε.

Lemma 5.3 (Céa's Lemma) *Let $u^\varepsilon \in H^1_0(\Omega^\varepsilon)$ solve (5.4), and $u^\varepsilon_h \in V^\varepsilon_h$ solve (5.25). Then there exists a constant c that is independent of ε such that*

$$\|u^\varepsilon - u^\varepsilon_h\|_{H^1(\Omega^\varepsilon)} \le c \inf_{v^\varepsilon_h \in V^\varepsilon_h} |u^\varepsilon - v^\varepsilon_h|_{H^1(\Omega^\varepsilon)}.$$

The goal now is to find a good approximation for u^ε in V^ε_h. We use the expansions for u^ε and of the basis functions as tools [137]. We consider now the expansion in K^ε_i of the basis functions λ_i defined in (5.24), corresponding to the node $(ih, 0) \in \Gamma \cap \mathcal{N}$. The expansion of λ_i em K^ε_{i-1} is similar. In K^ε_i,

$$\lambda_i(\mathbf{x}) = \frac{1}{h(d - \varepsilon z)} \big[(d + y - \varepsilon W - \varepsilon z)(x_{i+1} - x) + \varepsilon \theta_i(\mathbf{x}) + \varepsilon r_i(\mathbf{x}) \big], \qquad (5.30)$$

where $x_{i+1} = (i + 1)h$, and θ_i solve

$$-\Delta \theta_i = 0 \quad \text{in } K^\varepsilon_i,$$

$$\theta_i(x, y) = (x_{i+1} - x)\left[W(x, y) - \frac{zy}{d} \right] \quad \text{on } \partial K^\varepsilon_i \backslash \Gamma^\varepsilon_r, \qquad \theta_i = 0 \quad \text{on } \Gamma^\varepsilon_r.$$

Also

$$-\Delta r_i = 2\frac{\partial W}{\partial x} \quad \text{in } K^\varepsilon_i, \qquad r_i = 0 \quad \text{on } \partial K^\varepsilon_i.$$

Note that the expansion (5.27) in Ω^ε_r, and (5.30) are similar, since, in both cases, the low order terms involve $d + y - \varepsilon W - \varepsilon z$ times a function of x, in this case $\partial u^0/\partial y$ in (5.27), and $(x_{i+1} - x)/[h(d - \varepsilon z)]$ in (5.30).

The lemma below gives us an upper bound for the H^1 norms of r_i and θ_i.

Lemma 5.4 *Let r_i and θ_i as above. Then there exists a constant c independent of ε and h such that*

$$\|r_i\|_{H^1(K^\varepsilon_i)} \le c\varepsilon^{1/2}h^{1/2}, \qquad \|\theta_i\|_{H^1(K^\varepsilon_i)} \le ch.$$

Proof Classical estimates give $\|r_i\|_{H^1(K^\varepsilon_i)} \le \|W\|_{L^2(K^\varepsilon_i)}$. Arguing as in the proof of Lemma 5.2, we obtain the first estimate. It also follows from classical estimates that $\|\nabla \theta_i\|_{L^2(K^\varepsilon_i)} \le c\|\theta_i\|_{H^{1/2}(\partial K^\varepsilon_i \backslash \Gamma^\varepsilon_r)}$. Note that

$$\|\theta_i\|_{H^{1/2}(\partial K_i^\epsilon \setminus \Gamma_r^\epsilon)} \le h\|W\|_{H^{1/2}(\partial K_i^\epsilon \setminus \Gamma_r^\epsilon)} + hz.$$

From the exponential decay of W, and from interpolation and trace inequalities, we gather that $\|W\|_{H^{1/2}(\partial K_i^\epsilon \setminus \Gamma_r^\epsilon)} \le c$, and the result follows. $\square$

Theorem 5.1 *Let u^ϵ be a solution of (5.4), and $u_h^\epsilon \in V_h^\epsilon$ solution of (5.25). Assume (5.26), and that the trace of $\partial u^0/\partial y|_{\Omega_s}$ in Γ is in $H^2(\Gamma)$, where u^0 solve (5.8). Then there exists a constant c independent of ϵ and h such that*

$$\|u^\epsilon - u_h^\epsilon\|_{H^1(\Omega^\epsilon)} \le c(h + d^{3/2} + \epsilon h^{-1/2} + \epsilon^{3/2}h^{-1}).$$

Proof The proof is based on Lemma 5.3 and on the asymptotic expansions of u^ϵ and the functions in V_h^ϵ. Let $\Upsilon(\mathbf{x}) = (y - \epsilon W)\partial u^0/\partial y s(x, h)\chi_r^\epsilon$. From the triangle inequality and (5.29), we gather that

$$|u^\epsilon - v_h|_{H^1(\Omega^\epsilon)} \le |u^\epsilon - u^0 - u^1 + \Upsilon|_{H^1(\Omega^\epsilon)} + |u^0 + u^1 + \Upsilon - v_h|_{H^1(\Omega^\epsilon)}$$

$$\le cd^{3/2} + |u^0 + u^1 + \Upsilon - v_h|_{H^1(\Omega^\epsilon)} \tag{5.31}$$

for all $v_h \in V_h^\epsilon$. Choose v_h as $u_I(\mathbf{x}) = \sum_{\mathbf{x}_i \in \mathcal{N}}[u_0(\mathbf{x}_i) + u_1(\mathbf{x}_i)]\lambda_i(\mathbf{x})$. Hence, u_I is piecewise linear in Ω_s, continuous, and interpolates $u^0 + u^1$. Thus,

$$|u^0 + u^1 - u_I|_{H^1(\Omega_s)} \le ch. \tag{5.32}$$

In Ω_r^ϵ,

$$u_I(\mathbf{x}) = \left(d + y - \epsilon W - \epsilon z\right)I_h\left(\frac{\partial u^0}{\partial y}\right)(x) + R_I,$$

$$R_I(\mathbf{x}) = \epsilon h^{-1} \sum_{i=1}^{N} \frac{\partial u^0}{\partial y}(x_i, 0)[\theta_i(\mathbf{x}) + r_i(\mathbf{x})],$$

where $I_h(\partial u^0/\partial y)(\cdot)$ is the piecewise linear interpolation of $\partial u^0/\partial y(\cdot, 0)$ in $(0, 1)$. Let $e_I(x, y) = \partial u^0/\partial y(x, h) - I_h(\partial u^0/\partial y)(x)$ be the interpolation error. We shall need the estimates [54, 186]

$$\|e_I\|_{L^2(0,1)} + h\|\frac{\partial e_I}{\partial x}\|_{L^2(0,1)} + h\|e_I\|_{L^\infty(0,1)} + h^2\|\frac{\partial e_I}{\partial x}\|_{L^\infty(0,1)} \le ch^2\|\frac{\partial u^0}{\partial y}\|_{H^2(0,1)}.$$

Compute

$$|\Upsilon - u_I|_{H^1(\Omega_r^\epsilon)} \le |ye_I|_{H^1(\Omega_r^\epsilon)} + \epsilon|we_I|_{H^1(\Omega_r^\epsilon)} + |d - \epsilon z||I_h\left(\frac{\partial u^0}{\partial y}\right)|_{H^1(\Omega_r^\epsilon)} + |R_I|_{H^1(\Omega_r^\epsilon)}.$$

$$\tag{5.33}$$

Estimating the terms,

$$|ye_I|^2_{H^1(\Omega_r^\epsilon)} \le d \int_{\Omega_r^\epsilon} |\frac{\partial e_I}{\partial x}|^2 \, d\mathbf{x} + \int_{\Omega_r^\epsilon} |e_I|^2 \, d\mathbf{x} \le cd^2h^2 + cdh^4,$$

$$|we_I|^2_{H^1(\Omega_r^\epsilon)} \le \|e_I\|^2_{L^\infty(0,1)} |w|^2_{H^1(\Omega_r^\epsilon)} + \|\frac{\partial e_I}{\partial x}\|^2_{L^\infty(0,1)} |w|^2_{L^2(\Omega_r^\epsilon)}$$

$$\le c\epsilon^{-1}h^2 + c\epsilon, |d - \epsilon z|^2 |I_h\left(\frac{\partial u^0}{\partial y}\right)|^2_{H^1(\Omega_r^\epsilon)} \le cd^2 |I_h\left(\frac{\partial u^0}{\partial y}\right)|^2_{H^1(\Omega_r^\epsilon)}$$

$$\le cd^2 \int_{\Omega_r^\epsilon} |\frac{\partial I_h\left(\frac{\partial u^0}{\partial y}\right)}{\partial x}|^2 \, d\mathbf{x} \le cd^3.$$

From Lemma 5.4 we finally get

$$|R_I|^2_{H^1(K_i^\epsilon)} \le c\epsilon^2h^{-2}\left(\|\theta_i\|^2_{H^1(K_i^\epsilon)} + \|r_i\|^2_{H^1(K_i^\epsilon)}\right) \le c(\epsilon^2 + \epsilon^3 h^{-1}).$$

Adding over the elements, we obtain that $|R_I|_{H^1(K_i^\epsilon)} \le c(\epsilon h^{-1/2} + \epsilon^{3/2}h^{-1})$. Finally, from (5.33),

$$|\Upsilon - u_I|_{H^1(\Omega_r^\epsilon)} \le c(dh + d^{1/2}h^2 + \epsilon^{1/2}h + \epsilon^{3/2} + d^{3/2} + \epsilon h^{-1/2} + \epsilon^{3/2}h^{-1}). \tag{5.34}$$

The result follows from (5.26), (5.31), (5.32), and (5.34).   □

Remark 5.6 If $d \le c_1 h^{2/3}$, and $\epsilon \le c_2 h$ for given constants c_1, c_2, then

$$\|u^\epsilon - u_h^\epsilon\|_{H^1(\Omega^\epsilon)} \le ch + c\epsilon h^{-1/2}. \tag{5.35}$$

In practice, choosing d small reduces the computational costs associated with the basis functions computations.

In Theorem 5.1 and the estimate (5.35), the term $\epsilon h^{-1/2}$ is related to the choice of linear boundary conditions to define λ_i, that originates the spurious term θ_i in (5.30). This kind of *resonance* error is common in multiscale methods. For example, methods proposed to deal with PDEs with oscillatory coefficients, as described in Chap. 6 for instance, also present similar difficulties. There are several alternatives in the literature that try to deal with this [136–138, 188].

A possibility is to use *oversampling* [85, 134, 136, 138]. The goal is to free λ_i from being linear over the edges. Consider the element K_i^ϵ, and the expanded element

$$\hat{K}_i^\epsilon = \{(x,y) \in \Omega_r^\epsilon : ih - l < x < (i+1)h + l\},$$

for some $l > 0$. Consider the auxiliary multiscale functions ψ_i, ψ_{i+1} solutions of (5.24) in $\hat{K}_i^\epsilon$. We define then λ_i as

$$\lambda_i = c_i^1 \psi_i + c_i^2 \psi_{i+1},$$

and the constant c_i^1, c_i^2 are uniquely determined from the restrictions $\lambda_i(\mathbf{x}_j) = \delta_{ij}$ for $j = i, i+1$. The numerical solution is defined by (5.25), where again $V_h^\epsilon = \text{span}\{\lambda_i\}$.

To conclude the definition of the oversampling method, it is necessary to define l. Such constant must be large enough so that the boundary layer effect in K_i^ϵ is negligible. However, increasing l also increases the computational costs of computing basis functions. The sweet spot is $l = O(\varepsilon)$.

Note, however, that the method becomes nonconforming since λ_i is no longer continuous across edges, and thus $V_h^\epsilon \not\subset H^1(\Omega^\epsilon)$.

5.5 Conclusions

In this chapter we discussed how to develop an asymptotic expansion for PDEs posed in domains with rough boundaries. We also describe a finite element method of multiscale type to deal with such PDEs. Most of the previous works deal with this class problems using homogenization. For alternatives, see [90, 167, 171, 181].

The way to define the asymptotic expansion is by using a domain decomposition argument. This is an example of problem where the small parameter is in the domain itself. Other instances are in problems defined in rods, plates, and shells.

The numerical method described is quite general, in the sense that its definition does not rely on special characteristics of the wrinkles. Since the scheme is based on solution of local and independent problems, using parallel computation is trivial. The method is more efficient than fully resolving the domain with traditional schemes, but if the boundary is periodic, it is still cheaper to use the well-known wall-laws.

With respect to the error analysis, it is restricted to the periodic case, and the convergence rate in the $H^1(\Omega^\epsilon)$ is $h + \varepsilon h^{-1/2}$. The term $\epsilon h^{-1/2}$ is related to resonance errors, also present in multiscale methods for PDEs with oscillatory coefficients [85, 134, 136–138, 188].

Chapter 6
Partial Differential Equations with Oscillatory Coefficients

Abstract We present here some efficient techniques to approximate solutions of partial differential equations with oscillatory coefficients. We consider a simple one-dimensional case that still keeps most of the difficulties present in more sophisticated problems. We discuss three different approximation techniques: classical finite elements, homogenization, and Multiscale Finite Element methods (MsFEM). We show the advantages and pitfalls of each of the techniques, and present numerical results.

6.1 Introduction

Multiscale materials are problems of high interest, and present interesting mathematical and computational challenges. With the increase of computational power and the need of accurate modeling of new materials and complex systems, the area is receiving renewed attention.

It is easy to see that modeling nonlinear heterogeneous shells, for instance, is not an easy task. Other applications include simulation of flow in porous media, aquifer pollution, radioactive contamination, etc. [60]. In general, a practical way to model is to consider the material's macroscopic or homogenized information, and that is done by considering how the micro-structure influences the macroscopic behavior.

It is well known that the traditional finite element method with polynomials functions is not appropriate to solve PDEs of multiscale type. Indeed, traditional methods do not resolve the fine scales with reasonable computational costs, and might not be uniformly stable [186].

Different strategies that extend the traditional finite element method were developed to deal with the high costs associated with traditional methods. Babuška and Osborn [25, 26] proposed a general and flexible formulation in terms of which spaces should be used in the Galerkin method, but the choices are nontrivial and depend on the specifics of each problem.

A choice developed by Hou and Wu [136] was to build the finite element spaces by local solutions of the operator, and we concentrate our discussion on this type of method; see [80–88, 134, 135, 137, 138, 168, 201], and also [161, 162, 174, 188, 199] for different approaches.

© The Author(s) 2017
A.L. Madureira, *Numerical Methods and Analysis of Multiscale Problems*,
SpringerBriefs in Mathematics, DOI 10.1007/978-3-319-50866-5_6

We discuss here some numerical techniques that can be employed to approximate multiscale problems. We show the basic ideas behind the methods, in the simplest possible setting, but keeping in mind that the most interesting cases occur in higher dimensions.

In what follows we describe a model problem with periodic coefficients, with periodicity $\varepsilon > 0$, and develop a classical approximation based on *two-scale asymptotic expansions*. The derivation of such expansions has a lot in common with the matching asymptotics we develop in Chap. 2. Here however there is an extra ingredient; the terms of the expansion are assumed to depend on two variables, being periodic with respect to the second one with periodicity ε^{-1}. The first term of the expansion is the homogenized solution [8, 49, 56, 133].

We first investigate how well homogenization and, in Sect. 6.3, finite element method approximate the exact solution. A more efficient option is the Multiscale Finite Element Method [136, 137], described in Sect. 6.4. In Sect. 6.5 we add some final remarks.

6.2   Two-Scale Asymptotic Expansion

In this chapter, we focus on problem (1.5), but now with the extra assumption that $a^\varepsilon(x) = a(\varepsilon^{-1}x)$, where $a : \mathbb{R} \to \mathbb{R}$ is smooth and periodic with period one, i.e., $a(x + 1) = a(x)$, and that $\beta \geq a(x) \geq \alpha > 0$ for all $x \in \mathbb{R}$.

It is natural to ask the following:

1. Does the sequence of solutions u^ε converge to some function u^0 when $\varepsilon \to 0$?
2. The convergence holds in which sense, i.e., in which topology?
3. How fast is the convergence with respect to ε? Is it possible to estimate *rates* of convergence?
4. Finally, is there an equation that determines u^0? What are the properties of such equation?

We call u^0 the *homogenized solution*, and the corresponding problem as *homogenized problem or equation*.

At least formally, some answers are not that difficult to obtain. Using asymptotic expansion techniques, we assume that

$$u^\varepsilon(x) \sim u^0(x, \varepsilon^{-1}x) + \varepsilon u^1(x, \varepsilon^{-1}x) + \varepsilon^2 u^2(x, \varepsilon^{-1}x) + \cdots, \tag{6.1}$$

where

$$u^i : (0, 1) \times \mathbb{R} \to \mathbb{R}$$

$$(x, y) \mapsto u^i(x, y)$$

are to be determined. We impose an extra condition on u^i; they should be periodic with respect to y, with period ε. Although these conditions look ad hoc at a first sight, they allow the correct development of the expansion and that its terms are well defined. Moreover, such conditions are somehow intuitive, in the sense that since the microscale is periodic, it seems natural that we impose periodic conditions also in the y, also called "fast," variable. Indeed, the problem has two important scales, the macroscale described by $x \in (0, 1)$, and the microscale, described by the variable $y \in \mathbb{R}$.

Once again, the expansion (6.1) is, so far, formal, and should not be interpreted in the pointwise sense. It is natural to ask why the asymptotic expansion has such form. Actually, the expansion is an *ansatz*, an educated guess, and its validity can only be established a posteriori, showing that it approximates u^ε in some sense.

From the chain rule,

$$\frac{d}{dx}[u^i(x, \varepsilon^{-1}x)] = \left[\frac{\partial u^i}{\partial x}(x, y) + \varepsilon^{-1} \frac{\partial}{\partial y} u^i(x, y) \right]\Bigg|_{y=\varepsilon^{-1}x},$$

and similarly, from the operator definition

$$\mathscr{L}^\varepsilon v = \frac{d}{dx}\left(a^\varepsilon \frac{d}{dx} v \right), \tag{6.2}$$

for all $v : (0, 1) \to \mathbb{R}$ smooth enough, we have

$$(\mathscr{L}^\varepsilon u^i)(x) = \left\{ \varepsilon^{-2} \frac{\partial}{\partial y}\left[a(y)\frac{\partial}{\partial y} u^i(x, y) \right] + \varepsilon^{-1} \frac{\partial}{\partial x}\left[a(y)\frac{\partial}{\partial y} u^i(x, y) \right] \right.$$
$$\left. + \varepsilon^{-1} \frac{\partial}{\partial y}\left[a(y)\frac{\partial}{\partial x} u^i(x, y) \right] + \frac{\partial}{\partial x}\left[a(y)\frac{\partial}{\partial x} u^i(x, y) \right] \right\}\Bigg|_{y=\varepsilon^{-1}x}. \tag{6.3}$$

Formally replacing (6.1) in (6.3), we gather that

$$(\mathscr{L}^\varepsilon u^\varepsilon)(x) = \left\{ \varepsilon^{-2} \frac{\partial}{\partial y}\left[a(y)\frac{\partial}{\partial y} u^0(x, y) \right] + \varepsilon^{-1} \frac{\partial}{\partial y}\left[a(y)\frac{\partial}{\partial y} u^1(x, y) \right] \right.$$
$$+ \varepsilon^{-1} \frac{\partial}{\partial x}\left[a(y)\frac{\partial}{\partial y} u^0(x, y) \right] + \varepsilon^{-1} \frac{\partial}{\partial y}\left[a(y)\frac{\partial}{\partial x} u^0(x, y) \right]$$
$$+ \frac{\partial}{\partial y}\left[a(y)\frac{\partial}{\partial y} u^2(x, y) \right] + \frac{\partial}{\partial x}\left[a(y)\frac{\partial}{\partial y} u^1(x, y) \right] + \frac{\partial}{\partial y}\left[a(y)\frac{\partial}{\partial x} u^1(x, y) \right]$$
$$+ \frac{\partial}{\partial x}\left[a(y)\frac{\partial}{\partial x} u^0(x, y) \right] + \varepsilon \frac{\partial}{\partial y}\left[a(y)\frac{\partial}{\partial x} u^2(x, y) \right] + \varepsilon \frac{\partial}{\partial x}\left[a(y)\frac{\partial}{\partial y} u^2(x, y) \right]$$
$$\left. + \varepsilon \frac{\partial}{\partial x}\left[a(y)\frac{\partial}{\partial x} u^1(x, y) \right] + \varepsilon^2 \frac{\partial}{\partial x}\left[a(y)\frac{\partial}{\partial x} u^2(x, y) \right] + \cdots \right\}\Bigg|_{y=\varepsilon^{-1}x}. \tag{6.4}$$

Using (1.5) and (6.4), and gathering the ε^{-2} terms, we have

$$\frac{\partial}{\partial y}\left[a(y)\frac{\partial}{\partial y}u^0(x,y)\right]=0 \qquad \text{in } (0,1)\times(0,1). \tag{6.5}$$

Note that (6.5) holds for all $y \in (0,1)$, and not only for $y = \varepsilon^{-1}x$, as in (6.4). Therefore, it is posed in $(0,1)$ and parametrized by $x \in (0,1)$. Then, it follows from (6.5) that u^0 is independent of y, and henceforward we simply write $u^0(x)$. Grouping the ε^0 terms in (6.4),

$$\frac{\partial}{\partial y}\left[a(y)\frac{\partial}{\partial y}u^2(x,y)\right]+\frac{\partial}{\partial x}\left[a(y)\frac{\partial}{\partial y}u^1(x,y)\right]$$

$$+\frac{\partial}{\partial y}\left[a(y)\frac{\partial}{\partial x}u^1(x,y)\right]+\frac{\partial}{\partial x}\left[a(y)\frac{d}{dx}u^0(x)\right]=f \qquad \text{in } (0,1)\times\mathbb{R}. \tag{6.6}$$

Observe the compatibility condition of (6.6) that follows from integrating in $(0,1)$ with respect to y:

$$\frac{\partial}{\partial x}\int_0^1 a(y)\left[\frac{\partial}{\partial y}u^1(x,y)+\frac{d}{dx}u^0(x)\right]dy = f \qquad \text{in } (0,1). \tag{6.7}$$

To finally determine u^0, we collect the terms in ε^{-1} and conclude that

$$\frac{\partial}{\partial y}\left[a(y)\frac{\partial}{\partial y}u^1(x,y)\right]=-\frac{d}{dy}a(y)\frac{d}{dx}u^0(x) \qquad \text{in } (0,1)\times\mathbb{R}. \tag{6.8}$$

Solving

$$\frac{d}{dy}\left[a(y)\frac{d}{dy}H(y)\right]=-\frac{d}{dy}a(y) \quad \text{in } (0,1),$$

$$H(\cdot) \text{ periodic with period one}, \quad \int_0^1 H(y)\,dy = 0, \tag{6.9}$$

we gather that

$$u^1(x,y)=H(y)\frac{d}{dx}u^0(x) \tag{6.10}$$

solves (6.8). Replacing (6.7), we conclude that

$$A^*\frac{d^2}{dx^2}u^0 = f \quad \text{in } (0,1), \qquad u^0 = 0 \quad \text{over } \partial\Omega, \tag{6.11}$$

where the *effective diffusion* A^* is given by

$$A^* = \int_0^1 a(y)\left[\frac{d}{dy}H(y)+1\right]dy. \tag{6.12}$$

Remark 6.1 In one-dimensional problems, the effective diffusion is often written as $(\int_0^1 1/a(x)\,dx)^{-1}$, the harmonic mean value of a. Actually, this is the same as (6.12) since, from (6.9), $aH' + a = A^*$, and then $A^* = (\int_0^1 1/a(x)\,dx)^{-1}$ due to the periodicity of H. However, the formula employing the harmonic mean holds only in one-dimensional cases, whereas it is easy to generalize (6.9) and (6.12) for higher dimensions.

It is actually not too hard to obtain (6.11) directly, without resorting to asymptotic expansions. Also, the lower regularity assumption $a \in L^\infty(0, 1)$ suffices, besides periodicity [56].

The equations defined by (6.9) are called *cell problems*, and depend on the "local" behavior of $a(\cdot)$, and does not depend on ε or f. After solving (6.9), the problem (6.11) is well defined, and also independent of ε.

6.2.1 Justifying the Asymptotic Expansion

Although it is not possible to conclude from the above formal derivation that u^0 is indeed the limit of u^ε with respect to ε, there are techniques that are appropriate to prove the convergence, justifying (6.11). We mention, for instance, the method of oscillatory functions of Tartar, as well as the two-scale method of Nguetseng and Allaire [8, 56]. We develop here, simpler arguments, with a more limited range of application [134]. These arguments are enough for our goals, but require extra smoothness of the coefficients,

Let $z = u^\varepsilon - (u^0 + \varepsilon u^1 + \varepsilon^2 u^2)$. Note that both u^0 and u^1 are well defined by (6.11) and (6.10). Since the compatibility condition (6.7) holds, then there exists solution for (6.6). By direct substitution, we have that $\mathscr{L}^\varepsilon z = r$, where

$$
r(x) = \left\{ \varepsilon \frac{d}{dy}\left[a(y) \frac{d}{dx} u^2(x, y) \right] + \varepsilon \frac{d}{dx}\left[a(y) \frac{d}{dy} u^2(x, y) \right] \right.
$$
$$
\left. + \varepsilon \frac{d}{dx}\left[a(y) \frac{d}{dx} u^1(x, y) \right] + \varepsilon^2 \frac{d}{dx}\left[a(y) \frac{d}{dx} u^2(x, y) \right] \right\} \Bigg|_{y=\varepsilon^{-1}x}.
$$

Assuming that a and f are sufficiently smooth, we have that $\|r\|_{L^\infty(0,1)} \leq c\varepsilon$. Note that $z = -\varepsilon u^1 - \varepsilon^2 u^2$ on $\{0, 1\}$, and thus $\|z\|_{L^\infty(\partial\Omega)} \leq c\varepsilon$. From the maximum principle [112] for z, and from the definition of u^1 and u^2, we gather that

$$
\|u^\varepsilon - u^0\|_{L^\infty(0,1)} \leq \|u^\varepsilon - (u^0 + \varepsilon u^1 + \varepsilon^2 u^2)\|_{L^\infty(0,1)} + \|\varepsilon u^1\|_{L^\infty(0,1)}
$$
$$
+ \|\varepsilon^2 u^2\|_{L^\infty(0,1)} \leq c\varepsilon. \tag{6.13}
$$

Thus, the exact solution of (1.5) not only converges to the homogenized solution given by (6.11), but also the difference between them decreases linearly with ε in the $L^\infty(0, 1)$ norm. We have therefore answers for the questions posed in Sect. 6.2.

A convergence result as (6.13) can be obtained with weaker conditions. Let $\theta \in H^1(0, 1)$ be a weak solution of

$$-\frac{d}{dx}\left(a(x/\varepsilon)\frac{d\theta}{dx}(x)\right) = 0 \quad \text{in } (0, 1),$$

$$\theta(0) = u^1(0), \quad \theta(1) = u^1(1).$$

(6.14)

We have then the following result [137, Lemma 3.1, Corollary 3.2, Remark 3.3].

Theorem 6.1 *Let $a \in W^{1,p}(\mathbb{R})$, $p > 2$, periodic with period one. Assume that $f \in L^2(0, 1)$, and consider u^ε solution of (1.5). Let u^0, u^1, and θ be defined by (6.11), (6.10), and (6.14) respectively. Then, there exists a constant c independent of f and ε such that*

$$\|u^\varepsilon - u^0 - \varepsilon u^1 + \varepsilon\theta\|_{H^1(0,1)} \leq c\varepsilon\|u^0\|_{H^2(0,1)},$$

$$\|u^\varepsilon - u^0\|_{L^2(0,1)} \leq c\varepsilon\|u^0\|_{H^2(0,1)}.$$

Remark 6.2 In the above theorem $a \in W^{1,p}(\mathbb{R})$ simply means that both a and a' are in $L^p(\mathbb{R})$. Usually, in practical applications, $a(\cdot)$ is not smooth [189]. In composites, for instance, the coefficient is actually *discontinuous*, mimicking the inclusion of materials with different mechanical properties. However, the above result is important to obtain error estimates for the numerical methods that we propose further on.

6.2.2　An Example

Note that in the one-dimensional case, it is possible to obtain an analytic solution for (1.5):

$$u^\varepsilon(x) = -\int_0^x \frac{1}{a(\xi/\varepsilon)}\left(\int_0^\xi f(t)\,dt + c_0\right)d\xi,$$

$$c_0 = -\frac{1}{\int_0^1 \frac{1}{a(\xi/\varepsilon)}\,d\xi}\int_0^1 \frac{1}{a(\xi/\varepsilon)}\int_0^\xi f(t)\,dt\,d\xi.$$

In our examples, we consider

$$f(x) = 1, \quad a(x) = \frac{1}{2}(\beta - \alpha)(1 + \sin(2\pi x)) + \alpha, \quad \alpha = \frac{1}{2}, \quad \beta = \frac{5}{2}.$$

(6.15)

Consider the sequence of problems for $\varepsilon = 1/4$, $\varepsilon = 1/8$ and $\varepsilon = 1/16$. It is easy to see from Figs. 6.1, 6.2, and 6.3 that the frequency of oscillations of $a(\cdot/\varepsilon)$ grow as $\varepsilon \to 0$.

Fig. 6.1 Plots of $a(\cdot/\varepsilon)$ and the exact solution for $\varepsilon = 1/4$

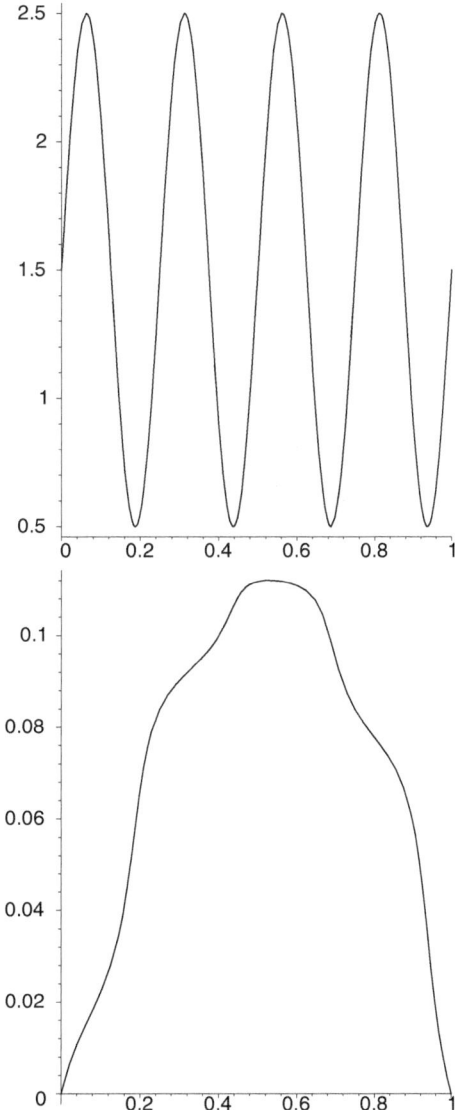

For higher dimensions, however, it is not possible to obtain analytic solutions in general. We seek then ways to find numerical approximations for (1.5). A possibility investigated is to employ homogenization technique, since, as $\varepsilon \to 0$, the exact solution converges to the homogenized solution. It is therefore reasonable to suspect that for small, but non-vanishing ε, both solutions are close to each other.

Assuming (6.15), we compare next both solutions. Consider the sequence of examples for $\varepsilon = 1/4$, $\varepsilon = 1/8$ and $\varepsilon = 1/16$. Note from Figs. 6.4 and 6.5 that, for ε small, the homogenized solution u^0 is a reasonably good approximation for the exact solution u^ε.

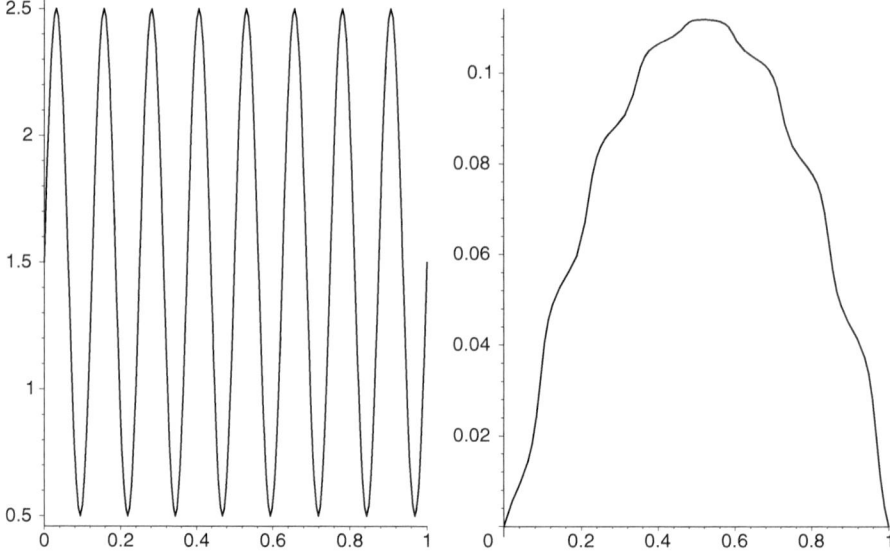

Fig. 6.2 Plots of $a(\cdot/\varepsilon)$ and the exact solution for $\varepsilon = 1/8$

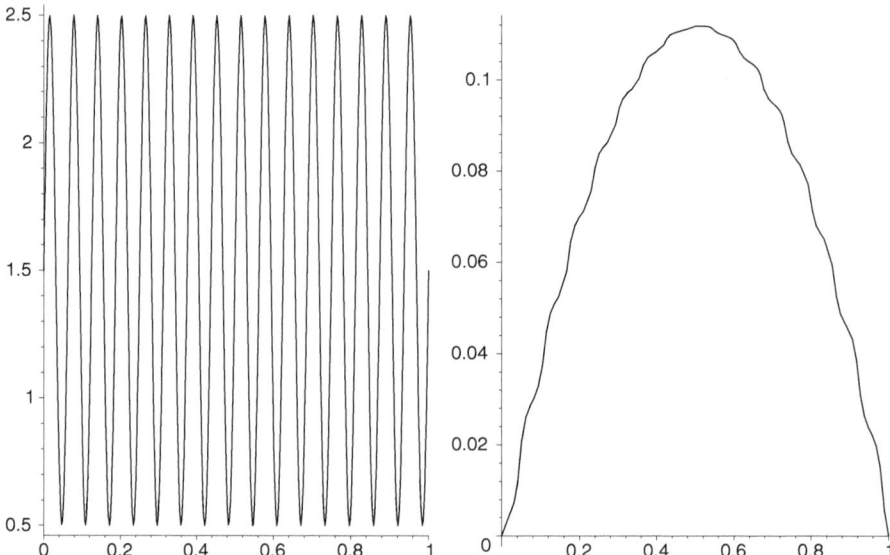

Fig. 6.3 Plots of $a(\cdot/\varepsilon)$ and the exact solution for $\varepsilon = 1/16$

Although they are very useful in several applications, homogenization techniques have their own limitations. For instance, its applicability is limited to values of ε small, as it is clear from Fig. 6.4. Further difficulties appear for more general cases, when $a(\cdot)$ is no longer periodic.

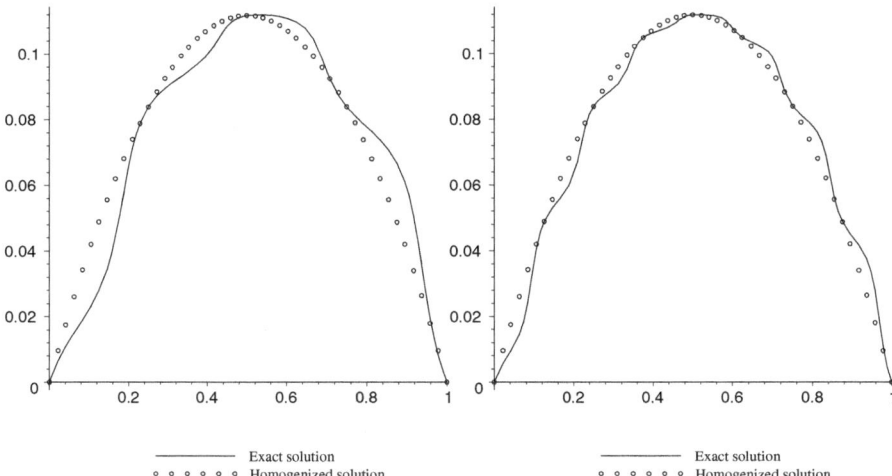

Fig. 6.4 Exact and homogenized solutions for $\varepsilon = 1/4$ and $\varepsilon = 1/8$

Fig. 6.5 Exact and homogenized solutions for $\varepsilon = 1/16$

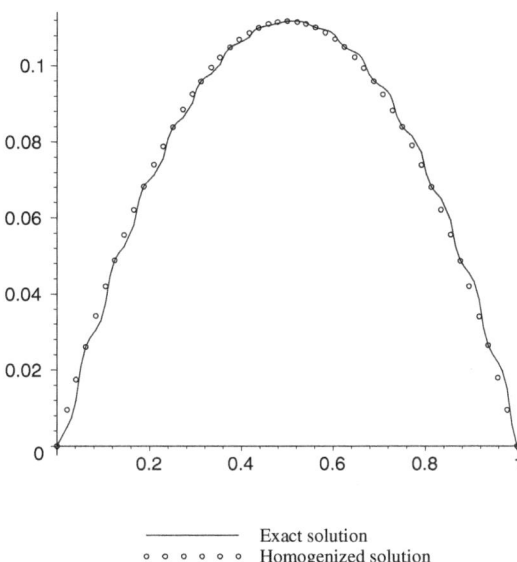

6.3   A Finite Element Method Discretization

The weak form of equation (1.5) is such that $u^\epsilon \in H_0^1(0, 1)$ solves (1.6). The existence and uniqueness of solutions for (1.6) follows from Poincaré's inequality (Lemma 1.1) and Lax–Milgram Lemma 1.2. If a and f are smooth, such solution also solves (1.5), and both formulations are equivalent.

Consider a discretization as described in Sect. 1.2, with V_h (cf. (1.12)) as

$$V_h = \{v_h \in H_0^1(0, 1) : v_h \text{ is linear in } (x_{j-1}, x_j) \text{ for } j = 1, \ldots, N + 1\}.$$

The finite element approximation for u^ϵ is $u_h \in V_h$ such that

$$\int_0^1 a(x/\varepsilon) \frac{du_h}{dx}(x) \frac{dv_h}{dx}(x) \, dx = \int_0^1 f(x)v_h(x) \, dx \quad \text{for all } v^h \in V_h. \tag{6.16}$$

Once again the Poincaré's inequality and the Lax–Milgram yield existence and uniqueness of solutions of (6.16).

Consider numerical approximations for (1.5), for a as in (6.15). For $\varepsilon = 1/4$ and $h = 1/32$, the finite element method approximates reasonably well the exact solution, as is shown in the left figure of Fig. 6.6. However, the approximation worsens as ε diminishes. See the graphs for $h = 1/32$ and $\varepsilon = 1/8$ at Fig. 6.6, and $\varepsilon = 1/16$ in Fig. 6.7 (on the left). The finite element approximation improves if we refine the mesh. For instance, considering $\varepsilon = 1/8$ with $h = 1/64$, we see some improvement in the approximation, as shown in Fig. 6.7.

What we want to remark is that the finite element method converges, but its convergence depends on ε. This can be an issue for higher dimensions, when the use of refined meshes is too expensive in terms of computations.

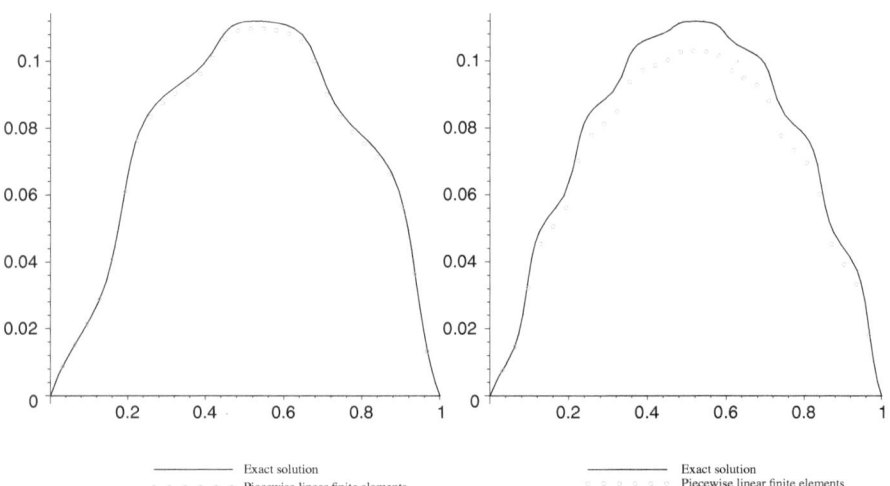

Fig. 6.6 Exact solution and its approximation for $h = 1/32$ with $\varepsilon = 1/4$ and $\varepsilon = 1/8$

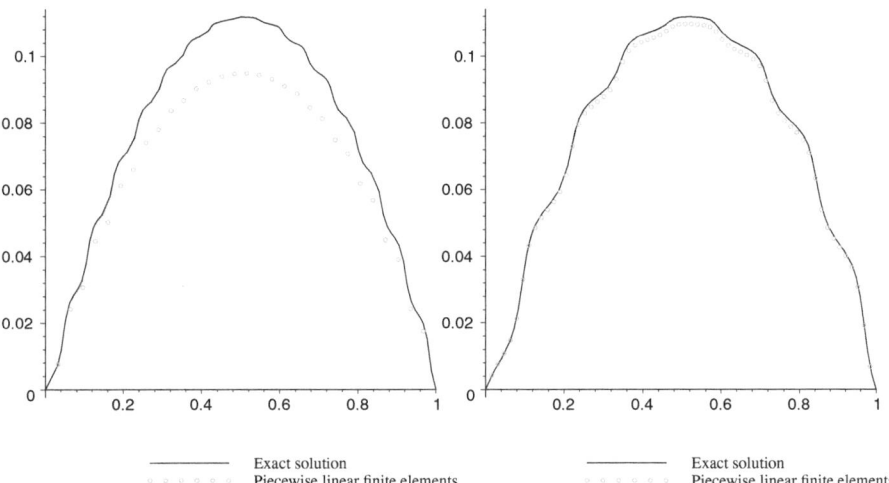

Fig. 6.7 *Left*: $\varepsilon = 1/16$ and $h = 1/32$. *Right*: $\varepsilon = 1/8$ and $h = 1/64$

6.3.1   What Goes Wrong?

To better understand why the classical finite element method does not work well, we develop an error analysis for this problem. The analysis is based on *Céa's Lemma* 1.3.

Lemma 6.1 (Céa's Lemma) *Let u^ε and u_h be solutions of* (1.5) *and* (6.16). *Then there exists a universal constant c such that*

$$\|u^\varepsilon - u_h\|_{H^1(0,1)} \leq c\frac{\beta}{\alpha}\|u^\varepsilon - v_h\|_{H^1(0,1)} \quad \textit{for all } v_h \in V_h. \tag{6.17}$$

Proof In our analysis, we use that $\beta \geq a(x) \geq \alpha > 0$. We start by investigating the continuity of the bilinear form $b(\cdot, \cdot)$. It follows from its definition that

$$b(u, v) \leq \beta \|u\|_{H^1(0,1)}\|v\|_{H^1(0,1)} \quad \text{for all } u, v \in H_0^1(0, 1). \tag{6.18}$$

Next, we estimate the coercivity:

$$b(v, v) \geq \alpha \int_0^1 \left(\frac{dv}{dx}\right)^2 dx \geq c\alpha\|v\|_{H^1(0,1)}^2 \quad \text{for all } v \in H_0^1(0, 1), \tag{6.19}$$

where we used Poincaré's inequality at the last step. The result follows from Céa's Lemma 1.3.   □

To obtain the final convergence result, it is enough to consider the interpolation result (1.15), that we rewrite here as

$$\|u^\varepsilon - I^h u^\varepsilon\|_{H^1(0,1)} \leq ch|u^\varepsilon|_{H^2(0,1)}, \tag{6.20}$$

where $I^h u^\epsilon = \sum_{j=1}^{N} u^\epsilon(x_j)\phi_j$ is the interpolation of u^ϵ in V_h. Then, substituting in (6.17) and using that

$$|u^\epsilon|_{H^2(0,1)} \leq \frac{c\|a'\|_{L^\infty(0,1)}}{\alpha^2 \epsilon} \|f\|_{L^2(0,1)}, \tag{6.21}$$

we have the following result.

Theorem 6.2 *Let $f \in L^2(0,1)$ and u^ϵ solution of (1.5). Then there exists a constant c independent of $\epsilon, f, \alpha,$ and β such that*

$$\|u^\epsilon - u^h\|_{H^1(0,1)} \leq c \frac{\beta \|a'\|_{L^\infty(0,1)}}{\alpha^3} \frac{h}{\epsilon} \|f\|_{L^2(0,1)}. \tag{6.22}$$

Interpreting the estimate of Theorem 6.2, we immediately gather that the method converges as $h \to 0$. Indeed, for a fixed ϵ, the approximation error goes to zero as h goes to zero. The problem is that the convergence with respect to h is not uniform in ϵ.

Thus, for small ϵ, unless the mesh is refined enough ($h \ll \epsilon$), the estimate (6.22) indicates that the error in the $H^1(0,1)$ norm is not small. That makes the classical finite element method almost useless for this kind of problem, and explains the bad numerical results indicated in the Figs. 6.6 and 6.7 (left plot).

6.4 Multiscale Finite Element Method

The proposal of Hou and Wu [136] to approximate PDEs with oscillatory coefficients is based on the idea of changing the basis functions of the finite element spaces. Instead of using piecewise polynomials, the multiscale finite element method employs functions that are local solutions (elementwise) of the original problem.

We present here the one-dimensional case. In almost all aspects, including the error analysis, the extension for two dimensions is natural. We remark at the end of this section when the generalization is nontrivial. Also, although the definition of the method in Sect. 6.4.1 is independent of the periodicity of $a(\cdot)$, the analysis presented in Sect. 6.4.2 is based on such property.

6.4.1 Definition of the Method

To define the method, we first construct the basis functions. Let ψ_i be such that

$$-\frac{d}{dx}\left(a^\epsilon(x)\frac{d\psi_i}{dx}(x)\right) = 0 \quad \text{in } \cup_{j=1}^{N+1}(x_{j-1}, x_j), \qquad \psi_i(x_j) = \delta_{ij} \tag{6.23}$$

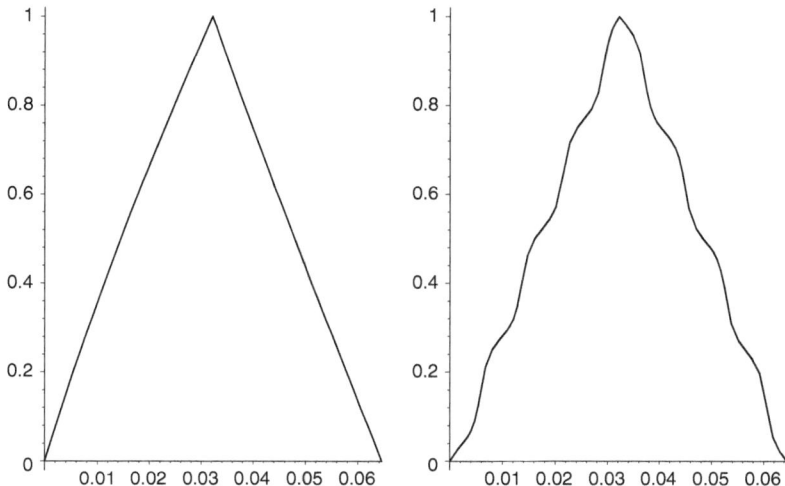

Fig. 6.8 For $h = 1/32$, plots of ψ_1 with $\varepsilon = 1/4$ and $\varepsilon = 1/128$

for $i = 1, \ldots, N$. Define then the multiscale finite element space by

$$V_0^{h,\varepsilon} = \text{span}\,\{\psi_1, \ldots, \psi_N\}.$$

Two typical basis functions are presented at Fig. 6.8, where we consider $a^\varepsilon(x) = a(x/\varepsilon)$ given by (6.15). On the left plot, the parameters are $\varepsilon = 1/4$ and $h = 1/32$. Note that the function looks like the usual finite element basis function. That is good news since in the case $h \ll \varepsilon$, and the traditional method works well; cf. the first comparison of Fig. 6.6. On the other hand, whenever $\varepsilon \ll h$, the basis function is oscillatory, as in the right plot of Fig. 6.8, with $\varepsilon = 1/128$ and $h = 1/32$.

We define then the multiscale finite element approximation $u^{h,\varepsilon} \in V_0^{h,\varepsilon}$ by

$$\int_0^1 a^\varepsilon(x) \frac{du^{h,\varepsilon}}{dx}(x) \frac{dv^{h,\varepsilon}}{dx}(x)\,dx = \int_0^1 f(x) v^{h,\varepsilon}(x)\,dx \quad \text{for all } v^{h,\varepsilon} \in V_0^{h,\varepsilon}. \tag{6.24}$$

In matrix formulation, if $u^{h,\varepsilon}(x) = \sum_{i=1}^N u_i^\varepsilon \psi_i(x)$, then $\mathbf{u}^\varepsilon = (u_1^\varepsilon, \ldots, u_N^\varepsilon)^T \in \mathbb{R}^N$ is such that

$$\mathbf{M}^\varepsilon \mathbf{u}^\varepsilon = \mathbf{f}^\varepsilon,$$

where the matrix $\mathbf{M}^\varepsilon = (M_{i,j}^\varepsilon) \in \mathbb{R}^{N \times N}$ and the vector $\mathbf{f}^\varepsilon = (f_1^\varepsilon, \ldots, f_N^\varepsilon)^T \in \mathbb{R}^N$ are given by

$$M_{i,j}^\varepsilon = \int_0^1 \left(a^\varepsilon(x) \frac{d\psi_i}{dx}(x) \frac{d\psi_j}{dx}(x) \right) dx, \qquad f_j^\varepsilon = \int_0^1 f(x) \psi_j(x)\,dx.$$

Fig. 6.9 Plot of u^ϵ and its numerical approximation by multiscale finite element method, with $\epsilon = 1/16$ and $h = 1/10$

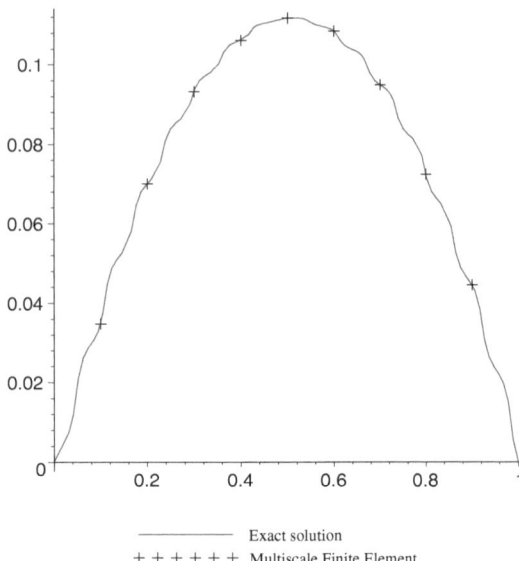

—————— Exact solution
+ + + + + + Multiscale Finite Element

As a numerical test, consider again (6.15) with $\epsilon = 1/16$ and $h = 1/10$. We see from Fig. 6.9 that the solution obtained from the multiscale method interpolates the exact solution at the nodes, as we remarked at Sect. 1.3.2.

6.4.2 Error Analysis

The error analysis considered in [137] is based on Céa's Lemma, as in Sect. 6.3.1.

In the classical finite element method, we find a function in V_h that "approximates well" u^ϵ and estimate the approximation error. Such function was the interpolation of u^ϵ, and the final result is obtained from Céa's Lemma 6.1.

The challenge is to find a good approximation for u^ϵ in $V_0^{h,\epsilon}$. This is performed considering separate cases, depending whether the mesh is refined enough with respect to ϵ or not. Actually, for one-dimensional problems a single case could be considered, but we maintain the separate analysis nonetheless. The reason is that, for higher dimensions, the error analysis gives different results depending whether $h \ll \epsilon$ or $\epsilon \ll h$.

Case I: $h \ll \epsilon$. When the mesh is refined enough, we obtain the following convergence result, which is, up to a constant, similar to of the Theorem 6.2. Thus, for refined meshes the multiscale and classical methods have similar performance. Note that in this analysis, there is no need to assume a^ϵ periodic.

Theorem 6.3 *Let $f \in L^2(0, 1)$, and let u^ϵ be the solution of (1.5). Then there exists a constant c that is independent of ϵ, f, α, and β such that*

$$\|u^\epsilon - u^{h,\epsilon}\|_{H^1(0,1)} \leq c\frac{\beta}{\alpha^2}h\|f\|_{L^2(0,1)}.$$

The above result follows from Céa's Lemma 6.1, and from the following interpolation result [137].

Lemma 6.2 *Let u^ϵ solution of (1.5), and let $I^{h,\epsilon}u^\epsilon = \sum_{j=1}^N u^\epsilon(x_j)\psi_j$ interpolating u^ϵ in $V_0^{h,\epsilon}$. Then, there exists a constant c independent of ϵ, f, α, and β such that*

$$\|u^\epsilon - I^{h,\epsilon}u^\epsilon\|_{H^1(0,1)} \leq c\frac{h}{\alpha}\|f\|_{L^2(0,1)}.$$

Proof Note that

$$\alpha|u^\epsilon - I^{h,\epsilon}u^\epsilon|^2_{H^1(x_{j-1},x_j)} \leq \int_{x_{j-1}}^{x_j} \frac{d}{dx}(u^\epsilon - I^{h,\epsilon}u^\epsilon)a^\epsilon(x)\frac{d}{dx}(u^\epsilon - I^{h,\epsilon}u^\epsilon)\,dx$$

$$= -\int_{x_{j-1}}^{x_j} (u^\epsilon - I^{h,\epsilon}u^\epsilon)\frac{d}{dx}\left[a^\epsilon(x)\frac{d}{dx}(u^\epsilon - I^{h,\epsilon}u^\epsilon)\right]dx$$

$$= -\int_{x_{j-1}}^{x_j} (u^\epsilon - I^{h,\epsilon}u^\epsilon)\frac{d}{dx}\left[a^\epsilon(x)\frac{d}{dx}u^\epsilon\right]dx$$

$$= \int_{x_{j-1}}^{x_j} (u^\epsilon - I^{h,\epsilon}u^\epsilon)f\,dx \leq \|u^\epsilon - I^{h,\epsilon}u^\epsilon\|_{L^2(x_{j-1},x_j)}\|f\|_{L^2(x_{j-1},x_j)}.$$

But the Poincaré's inequality yields

$$\|v\|_{L^2(x_{j-1},x_j)} \leq ch|v|_{H^1(x_{j-1},x_j)} \quad \text{for all } v \in H_0^1(x_{j-1}, x_j),$$

and then

$$\alpha|u^\epsilon - I^{h,\epsilon}u^\epsilon|^2_{H^1(x_{j-1},x_j)} \leq ch|u^\epsilon - I^{h,\epsilon}u^\epsilon|_{H^1(x_{j-1},x_j)}\|f\|_{L^2(x_{j-1},x_j)}.$$

Then,

$$|u^\epsilon - I^{h,\epsilon}u^\epsilon|_{H^1(x_{j-1},x_j)} \leq c\frac{h}{\alpha}\|f\|_{L^2(x_{j-1},x_j)}.$$

To find a global estimate, it is enough to add the above inequality over all the elements:

$$\|u^\epsilon - I^{h,\epsilon}u^\epsilon\|^2_{H^1(0,1)} \leq ch^2\sum_{j=1}^N \frac{1}{\alpha^2}\|f\|^2_{L^2(x_{j-1},x_j)} = c\frac{h^2}{\alpha^2}\|f\|^2_{L^2(0,1)},$$

and then the result follows.   □

Remark 6.3 The estimate of Theorem 6.3 holds only in one dimension. In two dimensions, the proof of Lemma 6.2 is different since $u^\epsilon - I^{h,\epsilon}u^\epsilon$ does not vanish on the element boundaries. That causes the error to behave as h/ϵ, which is no longer uniform with respect to ϵ.

Case II: $\epsilon \ll h$. Even when ϵ is small with respect to the mesh, and polynomial finite elements do not work well, the multiscale method does a good job in approximating the exact solution. Below we present an error estimate, under the periodicity assumption $a^\epsilon(x) = a(x/\epsilon)$.

Theorem 6.4 *Let $f \in L^2(0, 1)$, and let u^ϵ solve (1.5). Then there exists a constant c independent of ϵ and f such that*

$$\|u^\epsilon - u^{h,\epsilon}\|_{H^1(0,1)} \le c(\epsilon h^{-1/2} + h)\|f\|_{L^2(0,1)}.$$

To estimate the interpolation error of the present method we must find a function in $V_0^{h,\epsilon}$ that is a good approximation to u^ϵ, and then apply Céa's Lemma 6.1. Our candidate is $u_I \in V_0^{h,\epsilon}$ that interpolates u^0. Note that in **Case I** (when $h \ll \epsilon$), we considered a different interpolation.

To understand why this method works well even when $\epsilon \ll h$, we need to consider the first two terms of the asymptotic expansion of u^ϵ.

Hou et al. [137] noted that Theorem 6.1 holds for both the exact solution and the multiscale basis. Then, for $i = 1, \ldots, N$ the function ψ_i is approximated by $\psi_i^0 + \epsilon\psi_i^1 - \epsilon\theta_i$, where

$$-\frac{d^2}{dx^2}\psi_i^0 = 0 \quad \text{in } \cup_{j=1}^{N+1}(x_{j-1}, x_j), \qquad \psi_i(x_j) = \delta_{ij}$$

and $\psi_i^1 = H(x/\epsilon)d\psi_i^0/dx$. Finally,

$$-\frac{d}{dx}\left(a(x/\epsilon)\frac{d\theta_i}{dx}(x)\right) = 0 \quad \text{in } \cup_{j=1}^{N+1}(x_{j-1}, x_j), \qquad \theta_i(x_j) = \psi_i^1(x_j).$$

Remark 6.4 Note that $\psi_i^0 = \phi_i$ is simply the piecewise linear function.

As above, u_I can be approximated by $u_I^0 + \epsilon u_I^1 - \epsilon\theta_I$, where

$$u_I^0 = \sum_{i=1}^{N} u^0(x_i)\psi_i^0, \qquad u_I^1 = H(x/\epsilon)du_I^0/dx.$$

Moreover,

$$-\frac{d}{dx}\left(a(x/\epsilon)\frac{d\theta_I}{dx}(x)\right) = 0 \quad \text{in } \cup_{j=1}^{N+1}(x_{j-1}, x_j), \qquad \theta_I(x_j) = u_I^1(x_j).$$

It follows then that

$$
\begin{aligned}
\|u^\epsilon - u_I\|_{H^1(0,1)} \leq{}& \|u^\epsilon - u^0 - \epsilon u^1 + \epsilon\theta\|_{H^1(0,1)} + \|u^0 - u_I^0\|_{H^1(0,1)} \\
& + \epsilon\|u^1 - u_I^1\|_{H^1(0,1)} + \epsilon\|\theta\|_{H^1(0,1)} + \epsilon\|\theta_I\|_{H^1(0,1)} \\
& + \|u_I - u_I^0 - \epsilon u_I^1 + \epsilon\theta_I\|_{H^1(0,1)}.
\end{aligned}
\tag{6.25}
$$

The inequality

$$
\|u^\epsilon - u^0 - u^1 + \epsilon\theta\|_{H^1(0,1)} \leq C\epsilon\|u^0\|_{H^2(0,1)}
\tag{6.26}
$$

results from Theorem 6.1. Similarly,

$$
\|u_I - u_I^0 - u_I^1 + \epsilon\theta_I\|_{H^1(0,1)} \leq C\epsilon\|u^0\|_{H^2(0,1)}
\tag{6.27}
$$

follows from Theorem 6.1 and the estimate $\|u_I^0\|_{H^2(x_{j-1},x_j)} \leq C\|u^0\|_{H^2(x_{j-1},x_j)}$ (see details in [137]). To obtain

$$
\|u^0 - u_I^0\|_{H^1(0,1)} \leq Ch\|u^0\|_{H^2(0,1)},
\tag{6.28}
$$

it is enough to note that u_I^0 is the piecewise linear interpolation of u^0. Next, we use

$$
\begin{aligned}
\|u^1 - u_I^1\|_{H^1(x_{j-1},x_j)} ={}& \left\| H(\cdot/\epsilon)\frac{d(u^0 - u_I^0)}{dx} \right\|_{H^1(x_{j-1},x_j)} \\
\leq{}& \epsilon^{-1}\left\|\frac{dH}{dx}\right\|_{L^\infty(0,1)}\|u^0 - u_I^0\|_{H^1(x_{j-1},x_j)} \\
& + \|H\|_{L^\infty(0,1)}\|u^0 - u_I^0\|_{H^2(x_{j-1},x_j)} \\
\leq{}& C\epsilon^{-1}\|u^0 - u_I^0\|_{H^1(x_{j-1},x_j)} + C\|u^0\|_{H^2(x_{j-1},x_j)}.
\end{aligned}
$$

Summing the square of the above inequality from $j = 1$ to $j = N+1$ we gather that

$$
\|u^1 - u_I^1\|_{H^1(0,1)} \leq C(\epsilon^{-1}h + 1)\|u^0\|_{H^2(0,1)}.
\tag{6.29}
$$

Finally,

$$
\|\theta\|_{H^1(0,1)} \leq C(|u^1(0)| + |u^1(1)|) \leq C\|H\|_{L^\infty(0,1)}\left(\left|\frac{du^0}{dx}(0)\right| + \left|\frac{du^0}{dx}(1)\right|\right)
$$

$$
\leq C\|u^0\|_{H^2(0,1)},
\tag{6.30}
$$

and

$$\|\theta_I\|^2_{H^1(x_{j-1},x_j)} \leq Ch^{-1}(|u^1_I(x_{j-1})| + |u^1_I(x_j)|)^2$$

$$\leq Ch^{-1}\|H\|^2_{L^\infty(0,1)}\left(\left|\frac{du^0_I}{dx}(x_{j-1})\right| + \left|\frac{du^0_I}{dx}(x_j)\right|\right)^2$$

$$\leq Ch^{-1}\|u^0\|^2_{H^2(x_{j-1},x_j)}.$$

Adding the above from $j = 1$ to $j = N + 1$, we gather that

$$\|\theta_I\|_{H^1(0,1)} \leq Ch^{-1/2}\|u^0\|_{H^2(0,1)}. \tag{6.31}$$

Proof (of Theorem 6.4) To obtain the estimate, it is enough to use Lemma 6.1 and the inequalities (6.25)–(6.31), and the regularity result (6.21).   □

Remark 6.5 The convergence rate of Theorem 6.4 improves over what is stated in [169], where

$$\|u^\epsilon - u^{h,\epsilon}\|_{H^1(0,1)} \leq C_1h\|f\|_{L^2(0,1)} + C_2(\epsilon/h)^{1/2}.$$

The difference comes from the estimates of θ and θ_I, which are different in one- or two-dimensional problems.

6.4.3 An Extra Issue: High Contrast

An additional problem that can arise when dealing with heterogeneous media is the loss of coercivity—a particular case of high-contrast problem [45, 118]. Indeed, if α is too small, see (1.5), the problem becomes more difficult to deal with. Consider, for instance, the example given in (6.15), but with $\alpha = 0.01$, and $\epsilon = 1/8$. In Fig. 6.10 we plot $a(\cdot/\epsilon)$ and u^ϵ.

Even for ϵ small, the approximation for the homogenized solution is not satisfactory. Comparing the Figs. 6.2 and 6.11, we note the deterioration in the latter case.

Such deterioration is even more apparent when using piecewise finite elements, as shown in Figs. 6.7 (on the right) and 6.12. Note that this time the source of trouble is not the size of ϵ, but the magnitude of α. Indeed, even for ϵ relatively big, the classical finite element approximation fails. In Fig. 6.13 we show a numerical example for $\epsilon = 1/2$ and $h = 1/64$.

Once again, the deterioration was predicted by the error estimates. In Theorem 6.2, the constant is proportional to α^{-3}.

Finally, since the multiscale method interpolates the exact solution, it does not get worse as α gets smaller; see Fig. 6.14.

Fig. 6.10 Plots of $a(\cdot/\varepsilon)$ and
the exact solution for $\varepsilon = 1/8$
in a high contrast regime

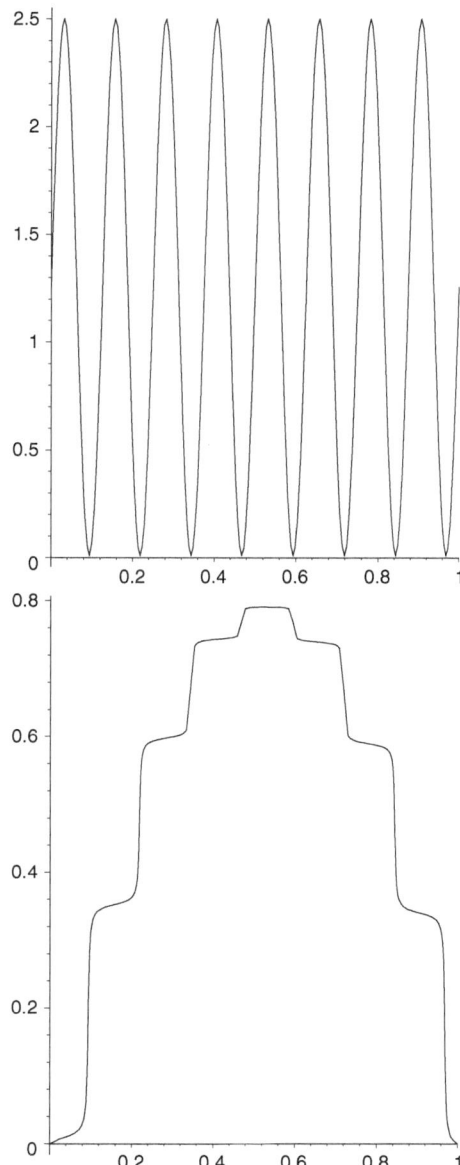

6.4.4   *Further Comments*

A major hurdle for the application of multiscale finite element methods in two
dimensions is that it is not clear what boundary conditions should be imposed for
the elementwise, local problems, see (6.23). In one dimension, such problem is not
present, since the edge reduces to a point.

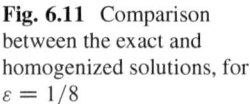

Fig. 6.11 Comparison
between the exact and
homogenized solutions, for
$\varepsilon = 1/8$

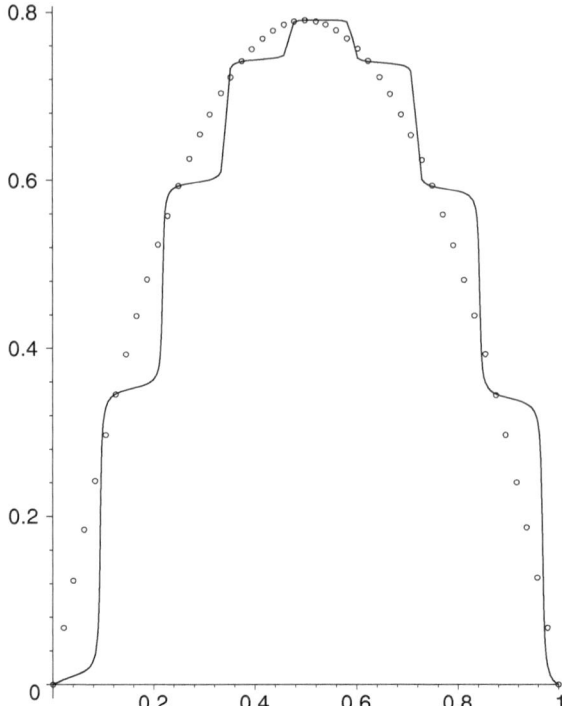

A first, and maybe natural idea is to impose the basis functions ψ_i to be linear over the edges. However this imposition originates spurious element-boundary layers (away from the real domain boundary), a phenomenon called *resonance*.

In [136, 137] it is proposed that the boundary conditions of the basis functions should also satisfy some sort of one-dimensional restriction of the PDE over the edges. This is of course an ad hoc idea since there is no "natural" restriction of a differential operator over an edge. However, it yields nice computational results. The convergence analysis in [137] is under the assumption that the basis functions are linear over the edges.

Another solution, proposed in [136] to deal with the resonance, analyzed in [85] is based on *oversampling*, making the method nonconforming.

Finally, Hou et al. [138] considers the idea of using a Petrov–Galerkin method, to decrease even further the effect of boundary layers. The use of Petrov–Galerkin methods to minimize the effect of the layers was also proposed in [102].

For nonlinear problems, the authors of [84, 87] propose and analyze a numerical homogenization technique. Through *G-convergence* concepts, they show that their schemes converge, up to subsequences. They are based on Petrov–Galerkin formulations on a nonlinear space of functions. It becomes clear that the method, named *nonlinear multiscale finite element method* (NMsFEM) is a generalization of the MsFEM of Hou and Wu [136].

Fig. 6.12 Plots of u^ε and its finite element approximation for $\varepsilon = 1/8$ and $h = 1/64$

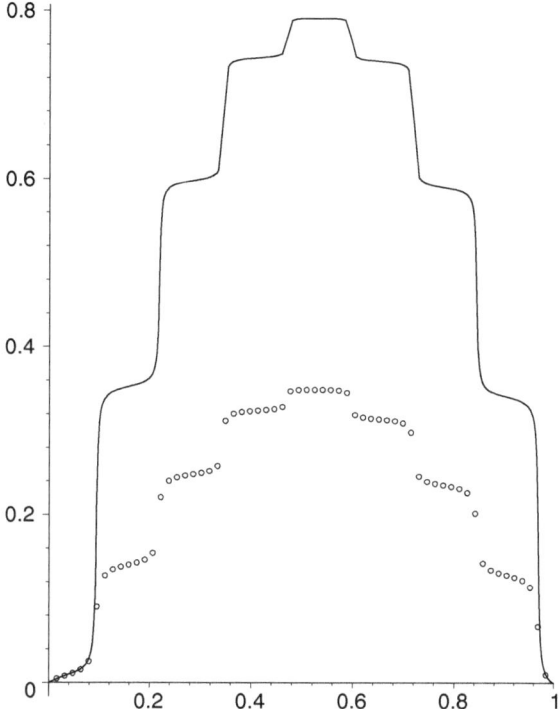

The use of exact or approximate solutions to construct the variational spaces as in [24–26, 136, 137] is not simple, since it might be hard to construct the correct space for a given problem. An interesting comparison showing how different choices of spaces influence the convergence rates for an one-dimensional advection problem is presented in [67].

The RFB formalism can actually guide the definition of the variational spaces. On the other hand, the choices imposed by the RFB introduce spurious components in the approximate solution. Sangalli [188] minimized such kind of effect by introducing *macro-bubbles*, making however the method much more expensive in terms of computations.

The HMM present interesting ideas, offering flexibility when defining the cell problems, in particular with costs that are independent of the small scales [73–79].

For problems with periodic coefficients, Versieux and Sarkis [196, 197] consider an efficient numerical scheme based on approximations of the terms in the asymptotic expansion of the solution. The final computational costs are also independent of ε.

Hybrid methods, as discussed in Sect. 1.3.5, lead to viable and powerful methods to deal with problems with rough coefficients. This particularly holds for the clever MHM method, which was successfully applied to a variety of multiscale problems; see [174] in particular.

Fig. 6.13 Plots of u^ϵ and its numerical approximation for $\varepsilon = 1/2$ and $h = 1/64$

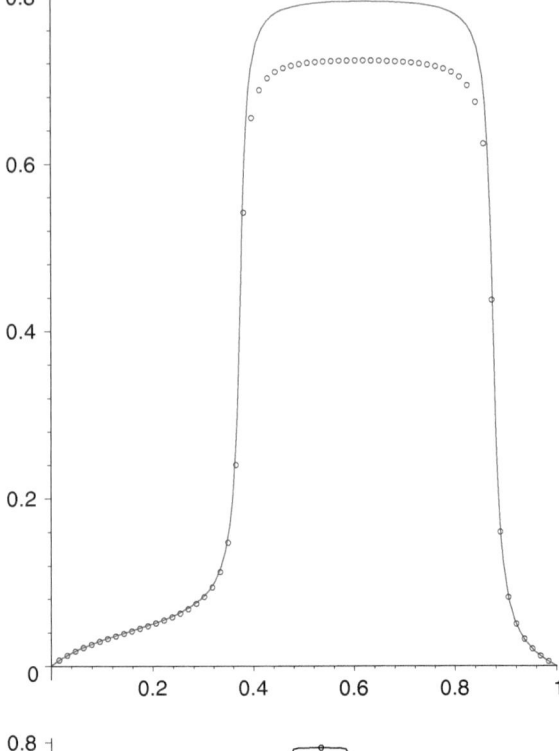

Fig. 6.14 Plots of u^ϵ and its multiscale finite element approximation, for $\varepsilon = 1/8$ and $h = 1/16$

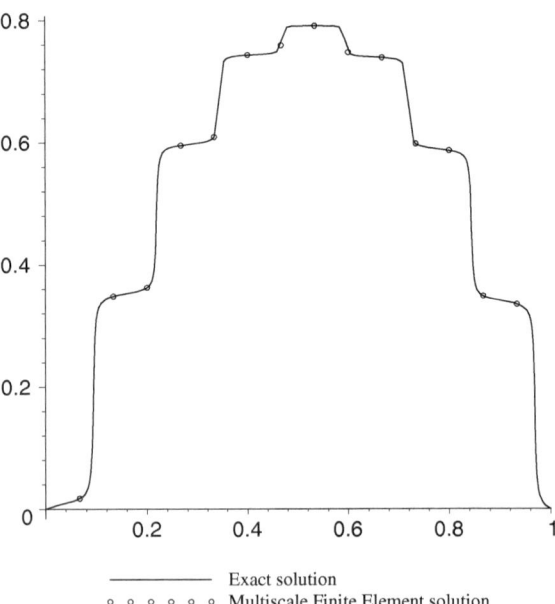

—————— Exact solution
∘ ∘ ∘ ∘ ∘ ∘ Multiscale Finite Element solution

6.5 Conclusions

Problems with oscillatory coefficients are of paramount importance since most physical media are heterogeneous by nature. These problems are complicate to deal with, and the one-dimensional setting considered here allows on to grasp at least what kind of difficulties one encounters when dealing with those PDEs.

We present a two-scale asymptotic expansion for the problem, and the way to compute the terms differs from the matching asymptotic expansions, defined for the previous problems. Furthermore, we introduce here the MsFEM in its simplest version, which was further developed in the literature to account for several physical aspects that are important for oil flow modeling [80, 81]. Our version, however, introduces the method and its proof of convergence in a way that preserves the main steps of the higher-dimensional counterpart while avoiding unpleasant technicalities.

References

1. Abboud, T., & Ammari, H. (1996). Diffraction at a curved grating: Approximation by an infinite plane grating. *Journal of Mathematical Analysis and Applications, 202*(3), 1076–1100. MR1408368 (98b:78028).
2. Abboud, T., & Ammari, H. (1996). Diffraction at a curved grating: TM and TE cases, homogenization. *Journal of Mathematical Analysis and Applications, 202*(3), 995–1026. MR1408364 (98b:78027).
3. Achdou, Y., Le Tallec, P., Valentin, F., & Pironneau, O. (1998). Constructing wall laws with domain decomposition or asymptotic expansion techniques. *Computational Methods in Applied Mechanical Engineering, 151*(1–2), 215–232. MR1625432 (99h:76027).
4. Achdou, Y., & Pironneau, O. (1995). Domain decomposition and wall laws. *Comptes Rendus de l'Académie des Sciences Paris Séries I Mathematics, 320*(5), 541–547 (English, with English and French summaries). MR1322334 (95k:76087).
5. Achdou, Y., Pironneau, O., & Valentin, F. (1998). Shape control versus boundary control. In *Équations aux dérivées partielles et applications* (pp. 1–18). Paris: Gauthier-Villars, Éd. Sci. Méd. Elsevier. MR1648212 (99j:49080).
6. Achdou, Y., Pironneau, O., & Valentin, F. (1998). Effective boundary conditions for laminar flows over periodic rough boundaries. *Journal of Computational Physics, 147*(1), 187–218. MR1657773 (99j:76086).
7. Alessandrini, S. M., Arnold, D. N., Falk, R. S., & Madureira, A. L. (1999). Derivation and justification of plate models by variational methods. In *Plates and shells (Québec, QC, 1996)*. CRM Proceedings of Lecture Notes (Vol. 21, pp. 1–20). Providence, RI: American Mathematical Society. MR1696513 (2000j:74055).
8. Allaire, G. (2002). Shape optimization by the homogenization method. In *Applied Mathematical Sciences* (Vol. 146). New York: Springer. MR1859696 (2002h:49001).
9. Allaire G., & Amar, M. (1999). Boundary layer tails in periodic homogenization. *ESAIM Control Optimization and Calculus of Variations, 4*, 209–243 (electronic) (English, with English and French summaries). MR1696289 (2000k:35019).
10. Amirat, Y., & Bodart, O. (2004). Numerical approximation of laminar flows over rough walls with sharp asperities. *Journal of Computational and Applied Mathematics, 164/165*, 25–38. MR2056866 (2004m:76118).
11. Amirat, Y., Bodart, O., De Maio, U., & Gaudiello, A. (2004). Asymptotic approximation of the solution of the Laplace equation in a domain with highly oscillating boundary. *SIAM Journal of Mathematical Analysis, 35*(6), 1598–1616 (electronic). MR2083791 (2005h:35064).

© The Author(s) 2017

A.L. Madureira, *Numerical Methods and Analysis of Multiscale Problems*, SpringerBriefs in Mathematics, DOI 10.1007/978-3-319-50866-5

12. Amirat, Y., Climent, B., Fernández-Cara, E., & Simon, J. (2001). The Stokes equations with Fourier boundary conditions on a wall with asperities. *Mathematical Methods in the Applied Science, 24*(5), 255–276. MR1818895 (2002b:76042).

13. Antonietti, P. F., Brezzi, F., & Marini, L. D. (2009). Bubble stabilization of discontinuous Galerkin methods. *Computational Methods in Applied Mechanical Engineering, 198*(21–26), 1651–1659. MR2517937.

14. Araya, R., Barrenechea, G. R., Franca, L. P., & Valentin, F. (2009). Stabilization arising from PGEM: A review and further developments. *Applied Numerical Mathematics, 59*(9), 2065–2081. doi:10.1016/j.apnum.2008.12.004. MR2532854.

15. Araya, R., Harder, C., Paredes, D., & Valentin, F. (2013). Multiscale hybrid-mixed method. *SIAM Journal of Numerical Analysis, 51*(6), 3505–3531. doi:10.1137/120888223. MR3143841.

16. Arbogast, T., Pencheva, G., Wheeler, M. F., & Yotov, I. (2007). A multiscale mortar mixed finite element method. *Multiscale Modelling and Simulation, 6*(1), 319–346. doi:10.1137/060662587. MR2306414.

17. Arnold, D. N., & Brezzi, F. (1985). Mixed and nonconforming finite element methods: Implementation, postprocessing and error estimates. *RAIRO Modélisation Mathématique et Analyse Numérique, 19*(1), 7–32 (English, with French summary). MR813687

18. Arnold, D. N., Brezzi, F., Cockburn, B., & Marini, L. D. (2001/2002). Unified analysis of discontinuous Galerkin methods for elliptic problems. *SIAM Journal of Numerical Analysis, 39*(5), 1749–1779. doi:10.1137/S0036142901384162. MR1885715 (2002k:65183).

19. Arnold, D. N., & Falk, R. S. (1996). Asymptotic analysis of the boundary layer for the Reissner-Mindlin plate model. *SIAM Journal of Mathematical Analysis, 27*(2), 486–514. MR1377485 (97i:73064).

20. Atkinson, K., & Han, W. (2005). *Theoretical numerical analysis. A functional analysis framework.* Texts in Applied Mathematics (2nd ed., Vol. 39). New York: Springer. MR2153422 (2006a:65001).

21. Auricchio, F., Bisegna, P., & Lovadina, C. (2001). Finite element approximation of piezoelectric plates. *International Journal for Numerical Methods in Engineering, 50*(6), 1469–1499. doi:10.1002/10970207(20010228)50:6ą 1469::AID-NME82£3.0.CO;2-I. MR1811534.

22. Axelsson, O., & Barker, V. A. (2001). *Finite element solution of boundary value problems: Theory and computation.* Classics in Applied Mathematics (Vol. 35). Philadelphia, PA: Society for Industrial and Applied Mathematics (SIAM); Reprint of the 1984 original. MR1856818 (2002g:65001)

23. Babuška, I. (1970/1971). Error-bounds for finite element method. *Numerical Mathematics, 16*, 322–333. MR0288971 (44#6166).

24. Babuška, I., Caloz, G., & Osborn, J. E. (1994). Special finite element methods for a class of second order elliptic problems with rough coefficients. *SIAM Journal of Numerical Analysis, 31*(4), 945–981. MR1286212 (95g:65146).

25. Babuška, I., & Osborn, J. E. (1983). Generalized finite element methods: Their performance and their relation to mixed methods. *SIAM Journal of Numerical Analysis, 20*(3), 510–536. MR701094 (84h:65076).

26. Babuška I., & Osborn, J. E. (1985). *Finite element methods for the solution of problems with rough input data (Oberwolfach, 1983).* Lecture Notes in Mathematics (Vol. 1121, pp. 1–18). Berlin: Springer. MR806382 (86m:65138).

27. Baer, S. M., Crook, S., Dur-E-Ahmad, M., & Jackiewicz, Z. (2009). Numerical solution of calcium-mediated dendritic branch model. *Journal of Computational and Applied Mathematics, 229*(2), 416–424. doi:10.1016/j.cam.2008.04.011.

28. Baiocchi, C., Brezzi, F., & Franca, L. P. (1993). Virtual bubbles and Galerkin-least-squares type methods (Ga.L.S.). *Computational Methods in Applied Mechanical Engineering, 105*(1), 125–141. doi:10.1016/0045-7825(93)90119-I. MR1222297 (94g:65058).

29. Barrenechea, G. R., Le Tallec, P., & Valentin, F. (2002). New wall laws for the unsteady incompressible Navier-Stokes equations on rough domains. *Mathematical Modelling and Numerical Analysis, 36*(2), 177–203. MR1906814 (2003k:76104).

30. Basson, A., & Gérard-Varet, D. (2008). Wall laws for fluid flows at a boundary with random roughness. *Communications on Pure and Applied Mathematics, 61*(7), 941–987. doi:10.1002/cpa.20237. MR2410410.
31. Boffi, D., Brezzi, F., & Fortin, M. (2013). *Mixed finite element methods and applications.* Springer Series in Computational Mathematics (Vol. 44). Heidelberg: Springer. MR3097958.
32. Bower, J. M., & Beeman, D. (2003). *The book of GENESIS: Exploring realistic neural models with the general neural simulation system.* Santa Clara, CA: TELOS.
33. Boyce, W. E., & DiPrima, R. C. (1965). *Elementary differential equations and boundary value problems.* New York/London/Sydney: Wiley. MR0179403 (31 #3651).
34. Brenner, S. C., & Scott, L. R. (2008). *The mathematical theory of finite element methods.* Texts in Applied Mathematics (3rd ed., Vol. 15). New York: Springer. MR2373954 (2008m:65001).
35. Bresch, D., & Milisic, V. (2006, November 6). *Higher order boundary layer corrector and wall laws derivation: A unified approach.* arXiv:math.AP/0611083v1.
36. Bressloff, P. C. (2001). Traveling fronts and wave propagation failure in an inhomogeneous neural network. *Physics D, 155*(1–2), 83–100. MR1837205 (2002d:92001).
37. Bressloff, P. C., Earnshaw, B. A., & Ward, M. J. (2008). Diffusion of protein receptors on a cylindrical dendritic membrane with partially absorbing traps. *SIAM Journal of Applied Mathematics, 68*(5), 1223–1246. MR2407121 (2009g:92017).
38. Brezis, H. (2011). *Functional analysis, Sobolev spaces and partial differential equations.* New York: Universitext, Springer. MR2759829.
39. Brezzi, F., Bristeau, M. O., Franca, L. P., Mallet, M., & Rogé, G. (1992). A relationship between stabilized finite element methods and the Galerkin method with bubble functions. *Computational Methods in Applied Mechanical Engineering, 96*(1), 117–129. doi:10.1016/0045-7825(92)90102-P. MR1159592 (92k:76056).
40. Brezzi, F., Franca, L. P., Hughes, T. J. R., & Russo, A. (1997). $b = \int g$, *Computational Methods in Applied Mechanical Engineering, 145*(3–4), 329–339. doi:10.1016/S0045-7825(96)01221-2. MR1456019 (98g:65086).
41. Brezzi, F., Franca, L. P., Hughes, T. J. R., & Russo, A. (1997). *Stabilization techniques and subgrid scales capturing.* The State of the Art in Numerical Analysis (York, 1996), Institute of Mathematics and Its Applications Conference Series (Vol. 63, pp. 391–406). New York: Oxford University Press. MR1628354 (99f:65162).
42. Brezzi, F., Franca, L. P., & Russo, A. (1998). Further considerations on residual-free bubbles for advective-diffusive equations. *Computational Methods in Applied Mechanical Engineering, 166*(1–2), 25–33. doi:10.1016/S0045-7825(98)00080-2. MR1660137 (99j:65197).
43. Brezzi, F., & Russo, A. (1994). Choosing bubbles for advection-diffusion problems. *Mathematical Models and Methods of Applied Science, 4*(4), 571–587. MR1291139 (95h:76079).
44. Brooks, A. N., & Hughes, T. J. R. (1982). Streamline upwind/Petrov-Galerkin formulations for convection dominated flows with particular emphasis on the incompressible Navier-Stokes equations. *Computational Methods in Applied Mechanical Engineering, 32*(1–3), 199–259. FENOMECH '81, Part I (Stuttgart, 1981). doi:10.1016/0045-7825(82)90071-8. MR679322 (83k:76005).
45. Burman, E., Guzman, J., Sanchez, M. A., & Sarkis, M. (2016). *Robust flux error estimation of Nitsche's method for high contrast interface problems.* arXiv:1602.00603v1 [math.NA].
46. Cai, D., Tao, L., Rangan, A. V., & McLaughlin, D. W. (2006). Kinetic theory for neuronal network dynamics. *Communications in Mathematical Science, 4*(1), 97–127. MR2204080 (2007a:82053).
47. Canic, S., Piccoli, B., Qiu, J.-M., & Ren, T. (2015). Runge-Kutta discontinuous Galerkin method for traffic flow model on networks. *Journal of Scientific Computing, 63*(1), 233–255. doi:10.1007/s10915-014-9896-z. MR3315275.
48. Carnevale, N., & Hines, M. L. (2006). *The NEURON book.* Cambridge: Cambridge University Press.

49. Chechkin, G. A., Piatnitski, A. L., & Shamaev, A. S. (2007). *Homogenization: Methods and applications*. Translations of Mathematical Monographs (Vol. 234). Providence, RI: American Mathematical Society. Translated from the 2007 Russian original by Tamara Rozhkovskaya. MR2337848.

50. Chen, C. (1995). *Asymptotic convergence rates for the Kirchhoff plate model*. Ph.D. Thesis, The Pennsylvania State University.

51. Chenais, D., & Paumier, J.-C. (1994). On the locking phenomenon for a class of elliptic problems. *Numerical Mathematics, 67*(4), 427–440. doi:10.1007/s002110050036. MR1274440.

52. Ciarlet, P. G. (1997). *Mathematical elasticity: Vol. II. Theory of plates*. Studies in Mathematics and Its Applications (Vol. 27). Amsterdam: North-Holland Publishing Co. MR1477663 (99e:73001).

53. Ciarlet, P. G. (2000). *Mathematical elasticity: Vol. III. Theory of shells*. Studies in Mathematics and Its Applications (Vol. 29). North-Holland Publishing Co., Amsterdam. MR1757535.

54. Ciarlet, P. G. (2002). *The finite element method for elliptic problems*. Classics in Applied Mathematics (Vol. 40). Philadelphia, PA: Society for Industrial and Applied Mathematics (SIAM). Reprint of the 1978 original [North-Holland, Amsterdam; MR0520174 (58 #25001)]. MR1930132.

55. Ciarlet, P. G. (2013). *Linear and nonlinear functional analysis with applications*. Philadelphia, PA: Society for Industrial and Applied Mathematics. MR3136903.

56. Cioranescu, D., & Donato, P. (1999). *An introduction to homogenization*. Oxford Lecture Series in Mathematics and Its Applications (Vol. 17). New York: The Clarendon Press/Oxford University Press. MR1765047 (2001j:35019).

57. Cioranescu, D., & Jean Paulin, J. St. (1999). *Homogenization of reticulated structures*. Applied Mathematical Sciences (Vol. 136). New York: Springer. MR1676922.

58. Cockburn, B., Dong, B., Guzmán, J., Restelli, M., & Sacco, R. (2009). A hybridizable discontinuous Galerkin method for steady-state convection-diffusion-reaction problems. *SIAM Journal of Scientific Computing, 31*(5), 3827–3846. doi:10.1137/080728810. MR2556564 (2010m:65216).

59. Cockburn, B., Gopalakrishnan, J., & Lazarov, R. (2009). Unified hybridization of discontinuous Galerkin, mixed, and continuous Galerkin methods for second order elliptic problems. *SIAM Journal of Numerical Analysis, 47*(2), 1319–1365. doi:10.1137/070706616. MR2485455.

60. Coutinho, A. L. G. A., Dias, C. M., Alves, J. L. D., Landau, L., Loula, A. F. D., Malta, S. M. C., et al. (2004). Stabilized methods and post-processing techniques for miscible displacements. *Computational Methods in Applied Mechanical Engineering, 193*(15–16), 1421–1436. MR2068902 (2005b:76083).

61. Coutinho, A. L. G. A., Franca, L. P., & Valentin, F. (2012). Numerical multiscale methods. *International Journal for Numerical Methods in Fluids, 70*(4), 403–419. doi:10.1002/fld.2727. MR2974524.

62. Cronin, J., & O'Malley, R. E. Jr. (Eds.). (1999). Analyzing multiscale phenomena using singular perturbation methods. *Proceedings of Symposia in Applied Mathematics* (Vol. 56). Providence, RI: American Mathematical Society. Dedicated to the memory of William A. Harris, Jr.; Papers from the American Mathematical Society Short Course held in Baltimore, MD, January 5–6, 1998. MR1722494.

63. Dauge, M., & Gruais, I. (1996). Asymptotics of arbitrary order for a thin elastic clamped plate, I: Optimal error estimates. *Asymptotic Analysis, 13*(2), 167–197. MR1413859 (98b:73022).

64. Dauge, M., & Gruais, I. (1998). Asymptotics of arbitrary order for a thin elastic clamped plate, II: Analysis of the boundary layer terms. *Asymptotic Analysis, 16*(2), 99–124. MR1612135.

65. Dautray, R., & Lions, J.-L. (1988). *Mathematical analysis and numerical methods for science and technology* (Vol. 2). Berlin: Springer. Functional and variational methods; With the collaboration of Michel Artola, Marc Authier, Philippe Bénilan, Michel Cessenat, Jean Michel Combes, Héléne Lanchon, Bertrand Mercier, Claude Wild and Claude Zuily; Translated from the French by Ian N. Sneddon. MR969367 (89m:00001).

66. Dayan, P., & Abbott, L. F. (2001). *Theoretical neuroscience: Computational and mathematical modeling of neural systems.* Computational Neuroscience. Cambridge, MA: MIT Press. MR1985615 (2004g:92008).

67. de Groen, P. P. N., & Hemker, P. W. (1979). Error bounds for exponentially fitted Galerkin methods applied to stiff two-point boundary value problems. In *Numerical analysis of singular perturbation problems* (Proc. Conf., Math. Inst., Catholic Univ., Nijmegen, 1978) (pp. 217–249). London: Academic Press. MR556520 (81a:65076).

68. De Schutter, E. (2001). Computational neuroscience: More math is needed to understand the human brain. In *Mathematics unlimited—2001 and beyond* (pp. 381–391). Berlin: Springer. MR1852166.

69. Di Pietro, D. A., & Ern, A. (2015). A hybrid high-order locking-free method for linear elasticity on general meshes. *Computational Methods in Applied Mechanical Engineering, 283,* 1–21. doi:10.1016/j.cma.2014.09.009. MR3283758.

70. do Carmo, M. P. (1976). *Differential geometry of curves and surfaces.* Englewood Cliffs, NJ: Prentice-Hall, Inc. Translated from the Portuguese. MR0394451 (52#15253).

71. Dostàl, Z., Horàk, D., & Kučera, R. (2006). Total FETI—an easier implementable variant of the FETI method for numerical solution of elliptic PDE. *Communications in Numerical Methods in Engineering, 22*(12), 1155–1162. doi:10.1002/cnm.881. MR2282408 (2007k:65177).

72. Dziuk, G., & Elliott, C. (2013). Finite element methods for surface PDEs. *Acta Numerica, 22,* 289–396. MR3038698.

73. E, W., & Engquist, B. (2003). Multiscale modeling and computation. *Notices of the American Mathematical Society, 50*(9), 1062–1070. MR2002752 (2004m:65163).

74. E, W., & Engquist, B. (2003). The heterogeneous multiscale methods. *Communications in Mathematical Science, 1*(1), 87–132. MR1979846 (2004b:35019).

75. E, W., & Engquist, B. (2005). The heterogeneous multi-scale method for homogenization problems. In *Multiscale methods in science and engineering.* Lecture Notes in Computational Science and Engineering (Vol. 44, pp. 89–110). Berlin: Springer. doi:10.1007/3-540-26444-2_4. MR2161708 (2006d:35017).

76. E, W., Engquist, B., & Huang, Z. (2003). Heterogeneous multiscale method: A general methodology for multiscale modeling. *Physical Review B, 67.* doi:10.1103/Phys-RevB. 67.092101.

77. E, W., Li, X., & Vanden-Eijnden, E. (2004). Some recent progress in multiscale modeling. In *Multiscale modelling and simulation.* Lecture Notes in Computational Science and Engineering (Vol. 39, pp. 3–21). Berlin: Springer, doi:10.1007/978-3-642-18756-8 1. MR2089950 (2005e:74049).

78. E, W., & Ming, P. (2004). Analysis of multiscale methods. *Journal of Computational Mathematics, 22*(2), 210–219. MR2058933 (2005d:65188).

79. E, W., Ming, P., & Zhang, P. (2005). Analysis of the heterogeneous multiscale method for elliptic homogenization problems. *Journal of the American Mathematical Society, 18*(1), 121–156 (electronic). MR2114818 (2005k:65246).

80. Efendiev, Y., Galvis, J., & Hou, T. Y. (2013). Generalized multiscale finite element methods (GMsFEM). *Journal of Computational Physics, 251,* 116–135. doi:10.1016/j.jcp.2013.04.045. MR3094911.

81. Efendiev, Y., Galvis, J., & Wu, X.-H. (2011). Multiscale finite element methods for high-contrast problems using local spectral basis functions. *Journal of Computational Physics, 230*(4), 937–955. doi:10.1016/j.jcp.2010.09.026. MR2753343.

82. Efendiev, Y., & Hou, T. Y. (2008). Multiscale computations for flow and transport in heterogeneous media. In *Quantum transport.* Lecture Notes in Mathematics (Vol. 1946, pp. 169–248). Berlin: Springer. MR2497877.

83. Efendiev, Y., & Hou, T. Y. (2009). *Multiscale finite element methods: Theory and applications.* Surveys and Tutorials in the Applied Mathematical Sciences (Vol. 4). New York: Springer. MR2477579.

84. Efendiev, Y., Hou, T., & Ginting, V. (2004). Multiscale finite element methods for nonlinear problems and their applications. *Communications in Mathematical Science, 2*(4), 553–589. MR2119929 (2005m:65265).

85. Efendiev, Y. R., Hou, T. Y., & Wu, X.-H. (2000). Convergence of a nonconforming multiscale finite element method. *SIAM Journal of Numerical Analysis, 37*(3), 888–910 (electronic). MR1740386 (2002a:65176).
86. Efendiev, Y., Lazarov, R., Moon, M., & Shi, K. (2015). A spectral multiscale hybridizable discontinuous Galerkin method for second order elliptic problems. *Computational Methods in Applied Mechanical Engineering, 292*, 243–256. doi:10.1016/j.cma.2014.09.036. MR3347248.
87. Efendiev, Y., & Pankov, A. (2003). Numerical homogenization of monotone elliptic operators. *Multiscale Modelling and Simulation, 2*(1), 62–79 (electronic). MR2044957 (2005a:65153).
88. Efendiev, Y. R., & Wu, X.-H. (2002). Multiscale finite element for problems with highly oscillatory coefficients. *Numerical Mathematics, 90*(3), 459–486. MR1884226 (2002m:65114).
89. Elfverson, D., Georgoulis, E. H., Målqvist, A., & Peterseim, D. (2013). Convergence of a discontinuous Galerkin multiscale method. *SIAM Journal of Numerical Analysis, 51*(6), 3351–3372. doi:10.1137/120900113. MR3141754.
90. Elfverson, D., Larson, M. G., & Maålqvist, A. (2015). *Multiscale methods for problems with complex geometry.* arXiv:1509.03991 [math.NA].
91. Ermentrout, B. (1998). Neural networks as spatial-temporal pattern-forming systems. *Reports on Progress in Physics, 61*, 353–430.
92. Ern, A., & Guermond, J.-L. (2004). *Theory and practice of finite elements.* Applied Mathematical Sciences (Vol. 159). New York: Springer. MR2050138 (2005d:65002).
93. Falk, R. S. (2008). Finite elements for the Reissner–Mindlin plate. In *Mixed finite elements, compatibility conditions, and applications.* Lecture Notes in Mathematics. New York: Springer.
94. Farhat, C., Harari, I., & Franca, L. P. (2001). The discontinuous enrichment method. *Computational Methods in Applied Mechanical Engineering, 190*(48), 6455–6479. doi:10.1016/S0045-7825(01)00232-8. MR1870426 (2002j:76083).
95. Farhat, C., & Roux, F.-X. (1991). A method of finite element tearing and interconnecting and its parallel solution algorithm. *International Journal for Numerical Methods in Engineering, 32*(6), 1205–1227. doi:10.1002/nme.1620320604.
96. Franca, L. P. & Dutra do Carmo, E. G. (1989). The Galerkin gradient least-squares method. *Computational Methods in Applied Mechanical Engineering, 74*(1), 41–54. doi:10.1016/0045-7825(89)90085-6. MR1017749 (90i:65195).
97. Franca, L. P., Farhat, C., Macedo, A. P., & Lesoinne, M. (1997). Residual-free bubbles for the Helmholtz equation. *International Journal for Numerical Methods in Engineering, 40*(21), 4003–4009. MR1475348.
98. Franca, L. P., Frey, S. L., & Hughes, T. J. R. (1992). Stabilized finite element methods, I: Application to the advective-diffusive model. *Computational Methods in Applied Mechanical Engineering, 95*(2), 253–276. doi:10.1016/0045-7825(92)90143-8. MR1155924 (92m:76089).
99. Franca, L. P., & Hwang, F.-N. (2002). Refining the submesh strategy in the two-level finite element method: Application to the advection-diffusion equation. *International Journal for Numerical Methods in Fluids, 39*(2), 161–187. doi:10.1002/fld.219. MR1903572.
100. Franca, L. P., & Madureira, A. L. (1993). Element diameter free stability parameters for stabilized methods applied to fluids. *Computational Methods in Applied Mechanical Engineering, 105*(3), 395–403. doi:10.1016/0045-7825(93)90065-6. MR1224304 (94g:76033).
101. Franca, L. P., Madureira, A. L., Tobiska, L., & Valentin, F. (2005). Convergence analysis of a multiscale finite element method for singularly perturbed problems. *Multiscale Modelling and Simulation, 4*(3), 839–866 (electronic). MR2203943 (2006k:65316).
102. Franca, L. P., Madureira, A. L., & Valentin, F. (2005).Towards multiscale functions: Enriching finite element spaces with local but not bubble-like functions. *Computational Methods in Applied Mechanical Engineering, 194*(27–29), 3006–3021. MR2142535 (2006a:65159).

103. Franca, L. P., Neslituk, A., & Stynes, M. (1998). On the stability of residual-free bubbles for convection-diffusion problems and their approximation by a two-level finite element method. *Computational Methods in Applied Mechanical Engineering, 166*(1–2), 35–49. doi:10.1016/S00457825(98)00081-4. MR1660133.
104. Franca, L. P., Ramalho, J. V. A., & Valentin, F. (2005). Multiscale and residual-free bubble functions for reaction advection-diffusion problems. *International Journal for Multiscale Engineering, 3,* 297–312.
105. Franca, L. P., Ramalho, J. V. A., & Valentin, F. (2006). Enriched finite element methods for unsteady reaction-diffusion problems. *Communications in Numerical Methods Engineering, 22*(6), 619–625. doi:10.1002/cnm.838. MR2235032.
106. Franca, L. P., & Russo, A. (1996). Approximation of the Stokes problem by residual-free macro bubbles. *East-West Journal of Numerical Mathematics, 4*(4), 265–278. MR1430240 (97i:76076).
107. Franca, L. P., & Russo, A. (1996). Deriving upwinding, mass lumping and selective reduced integration by residual-free bubbles. *Applied Mathematics Letters, 9*(5), 83–88. MR1415477 (97e:65121).
108. Franca, L. P., & Russo, A. (1997). Unlocking with residual-free bubbles. *Computational Methods in Applied Mechanical Engineering, 142*(3–4), 361–364. MR1442385 (98d:73060).
109. Franca, L. P., & Russo, A. (1997). Mass lumping emanating from residual-free bubbles. *Computational Methods in Applied Mechanical Engineering, 142*(3–4), 353–360. MR1442384 (98c:76064).
110. Galeão, A. C., Almeida, R. C., Malta, S. M. C., & Loula, A. F. D. (2004). Finite element analysis of convection dominated reaction-diffusion problems. *Applied Numerical Mathematics, 48*(2), 205–222. doi:10.1016/j.apnum.2003.10.002. MR2029331 (2004k:65219).
111. Gatica, G. N. (2014). *A simple introduction to the mixed finite element method: Theory and applications.* Springer Briefs in Mathematics. Cham: Springer. MR3157367.
112. Gilbarg, D., & Trudinger, N. S. (2001). *Elliptic partial differential equations of second order.* Classics in Mathematics. Berlin: Springer. Reprint of the 1998 edition. MR1814364 (2001k:35004).
113. Girault, V., & Raviart, P.-A. (1986). *Finite element methods for Navier-Stokes equations: Theory and algorithms.* Springer Series in Computational Mathematics (Vol. 5). Berlin: Springer. MR851383 (88b:65129).
114. Glowinski, R., & Wheeler, M. F. (1988). Domain decomposition and mixed finite element methods for elliptic problems. In *Partial differential equations (Paris, 1987)* (pp. 144–172). Philadelphia, PA: SIAM. MR972516 (90a:65237).
115. Golub, G. H., & Van Loan, C. F. (1983). *Matrix computations.* Johns Hopkins Series in the Mathematical Sciences (Vol. 3). Baltimore, MD: Johns Hopkins University Press. MR733103 (85h:65063).
116. Grisvard, P. (2011). *Elliptic problems in nonsmooth domains.* Classics in Applied Mathematics (Vol. 69). Philadelphia, PA: Society for Industrial and Applied Mathematics (SIAM). Reprint of the 1985 original [MR0775683]; With a foreword by Susanne C. Brenner. MR3396210.
117. Guzmán, J. (2006). Local analysis of discontinuous Galerkin methods applied to singularly perturbed problems. *Journal of Numerical Mathematics, 14*(1), 41–56. doi:10.1163/156939506776382157. MR2229818 (2007b:65122).
118. Guzman, J., Sanchez, M. A., & Sarkis, M. (2016). *A finite element method for high-contrast interface problems with error estimates independent of contrast.* arXiv:1507.03873v2.
119. Hackbusch, W. (2010). *Elliptic differential equations: Theory and numerical treatment.* Reprint of the 1992 English edition. Springer Series in Computational Mathematics (Vol. 18). Berlin: Springer. Translated from the 1986 corrected German edition by Regine Fadiman and Patrick D. F. Ion. MR2683186.
120. Hansel, D., Mato, G., Meunier, C., & Neltner, L. (1998). On numerical simulations of integrate-and-fire neural networks. *Neural Computation, 10*(2), 467–483.

121. Harari, I., & Hughes, T. J. R. (1994). Stabilized finite element methods for steady advection-diffusion with production. *Computational Methods in Applied Mechanical Engineering, 115*(1–2), 165–191. doi:10.1016/0045-7825(94)90193-7. MR1278815 (95a:76059).

122. Harder, C., Madureira, A.L., & Valentin, F. (2016). A hybrid-mixed method for elasticity. ESAIM Mathematical Modelling and Numerical Analysis, 50(2), 311–336. ISSN: 0764-583X; doi:10.1051/m2an/2015046. MR3482545.

123. Harder, C., Paredes, D., & Valentin, F. (2013). A family of multiscale hybrid-mixed finite element methods for the Darcy equation with rough coefficients. *Journal of Computational Physics, 245*, 107–130.

124. Harder, C., Paredes, D., & Valentin, F. (2015). On a multiscale hybrid-mixed method for advective-reactive dominated problems with heterogeneous coefficients. *Multiscale Modelling and Simulation, 13*(2), 491–518. doi:10.1137/130938499. MR3336297.

125. Haroske, D. D., & Triebel, H. (2008). *Distributions, Sobolev spaces, elliptic equations.* EMS Textbooks in Mathematics. Zürich: European Mathematical Society (EMS). MR2375667 (2009a:46003).

126. Henao, C. A. A., Coutinho, A. L. G. A., & Franca, L. P. (2010). A stabilized method for transient transport equations. *Computational Mechanics, 46*(1), 199–204. doi:10.1007/s00466-010-0465-5. MR2644409.

127. Herz, A. V. M., Gollisch, T., Machens, C. K., & Jaeger, D. (2006). Modeling single-neuron dynamics and computations: A balance of detail and abstraction. *Science, 314*(5796), 80–85. MR2253402 (2007d:92020).

128. Hesthaven, J. S., & Warburton, T. (2008). *Nodal discontinuous Galerkin methods: Algorithms, analysis, and applications.* Texts in Applied Mathematics (Vol. 54). New York: Springer. MR2372235 (2008k:65002).

129. Hines, M. (1984). Efficient computation of branched nerve equations. *International Journal of Bio-Medical Computing, 15*, 69–76.

130. Hines, M. L., & Carnevale, N. T. (1997). The NEURON simulation environment. *Neural Computation, 9*(6), 1179–1209.

131. Hines, M. L., Eichner, H., & Schürmann, F. (2008). Neuron splitting in compute-bound parallel network simulations enables runtime scaling with twice as many processors. *Journal of Computational Neuroscience, 25*(1), 203–210. doi:10.1007/s10827-007-0073-3.

132. Hines, M. L., Markram, H., & Schürmann, F. (2008). Fully implicit parallel simulation of single neurons. *Journal of Computational Neuroscience, 25*, 439–448. doi: 10.1007/s10827-008-0087-5.

133. Holmes, M. H. (2013). *Introduction to perturbation methods.* Texts in Applied Mathematics (2nd ed., Vol. 20). New York: Springer. MR2987304.

134. Hou, T. Y. (2003). Numerical approximations to multiscale solutions in partial differential equations. In *Frontiers in numerical analysis (Durham, 2002)* (pp. 241–301). MR2006969 (2004m:65219).

135. Hou, T. Y., & Liu, P. (2016). Optimal local multi-scale basis functions for linear elliptic equations with rough coefficient. *Discrete and Continuous Dynamical Systems, 36*(8), 4451–4476. ISSN:1078-0947; doi:10.3934/dcds.2016.36.4451. MR3479521.

136. Hou, T. Y., & Wu, X.-H. (1997). A multiscale finite element method for elliptic problems in composite materials and porous media. *Journal of Computational Physics, 134*(1), 169–189. MR1455261 (98e:73132).

137. Hou, T. Y., Wu, X.-H., & Cai, Z. (1999). Convergence of a multiscale finite element method for elliptic problems with rapidly oscillating coefficients. *Mathematics of Computation, 68*(227), 913–943. MR1642758 (99i:65126).

138. Hou, T. Y., Wu, X.-H., & Zhang, Y. (2004). Removing the cell resonance error in the multiscale finite element method via a Petrov-Galerkin formulation. *Communications in Mathematical Science, 2*(2), 185–205. MR2119937.

139. Hughes, T. J. R. (1978). A simple scheme for developing 'upwind' finite elements. *International Journal for Numerical Methods in Engineering, 12*, 1359–1365.

140. Hughes, T. J. R. (1987). *The finite element method: Linear static and dynamic finite element analysis*. Englewood Cliffs, NJ: Prentice Hall, Inc. With the collaboration of Robert M. Ferencz and Arthur M. Raefsky. MR1008473 (90i:65001).

141. Hughes, T. J. R. (1995). Multiscale phenomena: Green's functions, the Dirichlet-to-Neumann formulation, subgrid scale models, bubbles and the origins of stabilized methods. *Computational Methods in Applied Mechanical Engineering, 127*(1–4), 387–401. doi:10.1016/00457825(95)00844-9. MR1365381 (96h:65135).

142. Hughes, T. J. R., Feijóo, G. R., Mazzei, L., & Quincy, J.-B. (1998). The variational multiscale method—a paradigm for computational mechanics. *Computational Methods in Applied Mechanical Engineering, 166*(1–2), 3–24. doi:10.1016/S0045-7825(98)00079-6. MR1660141 (99m:65239).

143. Hughes, T. J. R., Franca, L. P., & Hulbert, G. M. (1989). A new finite element formulation for computational fluid dynamics, VIII: The Galerkin/least-squares method for advective-diffusive equations. *Computational Methods in Applied Mechanical Engineering 73*(2), 173–189. doi:10.1016/0045-7825(89)90111-4. MR1002621 (90h:76007).

144. Hughes, T. J. R., & Sangalli, G. (2007). Variational multiscale analysis: The fine-scale Green's function, projection, optimization, localization, and stabilized methods. *SIAM Journal of Numerical Analysis, 45*(2), 539–557. doi:10.1137/050645646. MR2300286 (2008c:65332).

145. Il'in, A. M. (1992). *Matching of asymptotic expansions of solutions of boundary value problems*. Translations of Mathematical Monographs (Vol. 102). Providence, RI: American Mathematical Society. Translated from the Russian by V. Minachin [V. V. Minakhin]. MR1182791.

146. Johnson, C. (1987). *Numerical solution of partial differential equations by the finite element method*. Cambridge: Cambridge University Press. MR925005 (89b:65003a).

147. Kellogg, R. B. (1992). *Notes on piecewise smooth elliptic boundary value problems*. College Park, MD: Institute for Physical Science and Technology.

148. Knabner, P., & Angermann, L. (2003). *Numerical methods for elliptic and parabolic partial differential equations*. Texts in Applied Mathematics (Vol. 44). New York: Springer. MR1988268 (2004j:65002).

149. Kozlov, V., Maz'ya, V., & Movchan, A. (1999). *Asymptotic analysis of fields in multi-structures*. Oxford Mathematical Monographs. New York: The Clarendon Press, Oxford University Press. Oxford Science Publications. MR1860617.

150. Kozlov, V. A., Maz'a, V. G., & Rossmann, J. (1997). *Elliptic boundary value problems in domains with point singularities*. Mathematical Surveys and Monographs (Vol. 52). Providence, RI: American Mathematical Society. MR1469972.

151. Kozubek, T., Vondrák, V., Menšík, M., Horák, D., Dostál, Z., Hapla, V., et al. (2013). Total FETI domain decomposition method and its massively parallel implementation. *Advances in Engineering Software, 60–61*, 14–22. doi:10.1016/j.advengsoft.2013.04.001.

152. Kreyszig, E. (1989). *Introductory functional analysis with applications*. Wiley Classics Library. New York: Wiley. MR992618.

153. Laing, C. R., Frewen, T. A., & Kevrekidis, I. G. (2007). Coarse-grained dynamics of an activity bump in a neural field model. *Nonlinearity, 20*, 2127–2146. doi:10.1088/0951-7715/20/9/007.

154. Lax, P. D. (2002). *Functional analysis*. Pure and Applied Mathematics (New York). New York: Wiley Interscience. MR1892228.

155. Lions, J.-L., & Magenes, E. (1972). *Non-homogeneous boundary value problems and applications* (Vol. I). New York: Springer. Translated from the French by P. Kenneth; Die Grundlehren der mathematischen Wissenschaften, Band 181. MR0350177 (50#2670).

156. Madureira, A. L. (2009). A multiscale finite element method for partial differential equations posed in domains with rough boundaries. *Mathematics of Computation, 78*(265), 25–34. MR2448695.

157. Madureira, A. L. (2015). Abstract multiscale-hybrid-mixed methods. *Calcolo, 52*(4), 543–557. MR3421669.

158. Madureira, A. L., Madureira, D. Q. M., & Pinheiro, P. O. (2012). A multiscale numerical method for the heterogeneous cable equation. *Neurocomputing, 77*(1), 48–57.
159. Madureira, A., & Valentin, F. (2002). Analysis of curvature influence on effective boundary conditions. *Comptes Rendus de L'Academie des Sciences Paris, 335*(5), 499–504 (English, with English and French summaries). MR1937121 (2003j:35054).
160. Madureira, A. L., & Valentin, F. (2006/2007). Asymptotics of the Poisson problem in do mains with curved rough boundaries. *SIAM Journal of Mathematical Analysis, 38*(5), 1450–1473 (electronic). MR2286014.
161. Målqvist, A. (2011). Multiscale methods for elliptic problems. *Multiscale Modelling and Simulation, 9*(3), 1064–1086. doi:10.1137/090775592. MR2831590 (2012j:65419).
162. Målqvist, A., & Peterseim, D. (2014). Localization of elliptic multiscale problems. *Mathematics of Computation, 83*(290), 2583–2603. doi:10.1090/S0025-5718-2014-02868-8. MR3246801.
163. Mclaughlin, D., Shapley, R., & Shelley, M. (2003). Large-scale modeling of the primary visual cortex: Influence of cortical architecture upon neuronal response. *Journal of Physiology-Paris, 97*, 237–252.
164. McLean, W. (2000). *Strongly elliptic systems and boundary integral equations*. Cambridge: Cambridge University Press. MR1742312.
165. Meunier, C., & Lamotte d'Incamps, B. (2008). Extending cable theory to heterogeneous dendrites. *Neural Computing, 20*(7), 1732–1775. MR2417105 (2009e:92021).
166. Meunier, C., & Segev, I. (2002). Playing the devil's advocate: Is the Hodgkin-Huxley model useful? *Trends in Neuroscience, 25*(11), 558–563.
167. Ming, P., & Xu, X. (2016). A multiscale finite element method for oscillating Neumann problem on rough domain. *Multiscale Modeling and Simulation, 14*(4), 1276–1300. ISSN:1540-3459; doi:10.1137/15M1044709. MR3554877.
168. Ming, P., & Yue, X. (2006). Numerical methods for multiscale elliptic problems. *Journal of Computational Physics, 214*(1), 421–445. MR2208685 (2006j:65359).
169. Moskow, S., & Vogelius, M. (1997). First-order corrections to the homogenised eigenvalues of a periodic composite medium: A convergence proof. *Proceedings of the Royal Society of Edinburgh Section A, 127*(6), 1263–1299. MR1489436 (99g:35018).
170. Nečas, J. (1967). *Les méthodes directes en théorie des équations elliptiques*. Masson et Cie, Éditeurs, Paris; Academia, Éditeurs, Prague (French). MR0227584 (37 #3168).
171. Nicaise, S., & Sauter, S. A. (2006) Efficient numerical solution of Neumann problems on complicated domains. *Calcolo, 43*(2), 95–120. MR2245226 (2007c:65108).
172. Novotny, A. A., & Sokołowski, J. (2013). *Topological derivatives in shape optimization*. Interaction of Mechanics and Mathematics. Heidelberg: Springer. MR3013681.
173. Omurtag, A., Knight, B. W., & Sirocich, L. (2000). On the simulation of large populations of neurons. *Journal of Computational Neuroscience, 8*(8), 51–63.
174. Paredes, D., Valentin, F., & Versieux, H.M. (2017). On the robustness of multiscale hybrid-mixed methods. *Mathematics of Computation, 86*(304), 525–548. ISSN: 0025-5718; doi:10.1090/mcom/3108. MR3584539.
175. Peterseim, D., & Sauter, S. A. (2011). Finite element methods for the Stokes problem on complicated domains. *Computational Methods in Applied Mechanical Engineering, 200* (33–36), 2611–2623. doi:10.1016/j.cma.2011.04.017. MR2812028.
176. Pian, T., & Tong, P. (1969). Basis of finite element methods for solid continua. *International Journal for Numerical Methods in Engineering, 1*, 3–28.
177. Pokornyi, Y. V., & Borovskikh, A. V. (2004). Differential equations on networks (geometric graphs). *Journal of Mathematical Science (N. Y.), 119*(6), 691–718. doi:10.1023/B:JOTH.0000012752.77290.fa. MR2070600.
178. Quarteroni, A., & Valli, A. (1994). *Numerical approximation of partial differential equations*. Springer Series in Computational Mathematics (Vol. 23). Berlin: Springer. MR1299729 (95i:65005).

179. Rangan, A. V., & Cai, D. (2007). Fast numerical methods for simulating large-scale integrate-and-fire neuronal networks. *Journal of Computational Neuroscience, 22*, 81–100. doi:10.1007/s10827-006-8526-7.
180. Raviart, P.-A., & Thomas, J. M. (1977). Primal hybrid finite element methods for 2nd order elliptic equations. *Mathematics of Computation, 31*(138), 391–413. MR0431752 (55 #4747).
181. Rech, M., Sauter, S., & Smolianski, A. (2006). Two-scale composite finite element method for Dirichlet problems on complicated domains. *Numerical Mathematics, 102*(4), 681–708. MR2207285 (2006m:65283).
182. Rempe, M. J., & Chopp, D. L. (2006). A predictor-corrector algorithm for reaction-diffusion equations associated with neural activity on branched structures. *SIAM Journal of Scientific Computing, 28*(6), 2139–2161 (electronic). MR2272255 (2008f:65148).
183. Rempe, M. J., Spruston, N., Kath, W. L., & Chopp, D. L. (2008). Compartmental neural simulations with spatial adaptivity. *Journal of Computational Neuroscience, 25*, 465–480. doi:10.1007/s10827-008-0089-3.
184. Rochinha, F. A., & Madureira, A. L. (2004). *Modelagem Multiescala em Materiais e Estruturas*. Notas em Matemática Aplicada (Vol. 12). São Carlos: SBMAC.
185. Roggensack, A. (2013). A kinetic scheme for the one-dimensional open channel flow equations with applications on networks. *Calcolo, 50*(4), 255–282. doi:10.1007/s10092-012-0066-0. MR3118265.
186. Roos, H.-G., Stynes, M., & Tobiska, L. (2008). *Robust numerical methods for singularly perturbed differential equations. Convection-diffusion-reaction and flow problems*. Springer Series in Computational Mathematics (2nd ed., Vol. 24). Berlin: Springer. MR2454024 (2009f:65002).
187. Rubin, J., & Wechselberger, M. (2008). The selection of mixed-mode oscillations in a Hodgkin-Huxley model with multiple timescales. *Chaos, 18*(1), 015105, 12. MR2404661 (2009a:37194).
188. Sangalli, G. (2003). Capturing small scales in elliptic problems using a residual-free bubbles finite element method. *Multiscale Modelling and Simulation, 1*(3), 485–503 (electronic). MR2030161 (2004m:65202).
189. Sarkis, M., & Versieux, H. (2008). Convergence analysis for the numerical boundary corrector for elliptic equations with rapidly oscillating coefficients. *SIAM Journal of Numerical Analysis, 46*(2), 545–576. MR2383203.
190. Shelley, M. J., & Mclaughlin, D. W. (2002). Coarse-grained reduction and analysis of a network model of cortical response, I: Drifting grating stimuli. *Journal of Computational Neuroscience, 12*, 97–122.
191. Shelley, M. J., & Tao, L. (2001). Efficient and accurate time-stepping schemes for integrate-and-fire neuronal networks. *Journal of Computational Neuroscience, 11*, 111–119.
192. Toselli, A., & Widlund, O. (2005). *Domain decomposition methods—algorithms and theory*. Springer Series in Computational Mathematics (Vol. 34). Berlin: Springer. MR2104179 (2005g:65006).
193. Tuckwell, H. C. (1988). *Introduction to theoretical neurobiology. Vol. 1: Linear cable theory and dendritic structure*. Cambridge Studies in Mathematical Biology (Vol. 8). Cambridge: Cambridge University Press. MR947344 (90a:92003a).
194. Tuckwell, H. C. (1988). *Introduction to theoretical neurobiology. Vol. 2: Nonlinear and stochastic theories*. Cambridge Studies in Mathematical Biology (Vol. 8). Cambridge: Cambridge University Press. MR947345 (90a:92003b).
195. Verhulst, F. (2005). *Methods and applications of singular perturbations. Boundary layers and multiple timescale dynamics*. Texts in Applied Mathematics (Vol. 50). New York: Springer. MR2148856.
196. Versieux, H. M., & Sarkis, M. (2006). Numerical boundary corrector for elliptic equations with rapidly oscillating periodic coefficients. *Communications in Numerical Methods in Engineering, 22*(6), 577–589. MR2235030 (2007d:65117).

197. Versieux, H., & Sarkis, M. (2007). A three-scale finite element method for elliptic equations with rapidly oscillating periodic coefficients. *Domain decomposition methods in science and engineering XVI*. Lecture Notes in Computational Science and Engineering (Vol. 55, pp. 763–770). Berlin: Springer. MR2334173.
198. Walker, S. W. (2015). *The shapes of things: A Practical guide to differential geometry and the shape derivative*. Advances in design and control (Vol. 28). Philadelphia, PA: Society for Industrial and Applied Mathematics (SIAM).
199. Wang, W., Guzmán, J., & Shu, C.-W. (2011). The multiscale discontinuous Galerkin method for solving a class of second order elliptic problems with rough coefficients. *International Journal of Numerical Analysis and Modeling, 8*(1), 28–47. MR2740478 (2012a:65346).
200. Zhuang, Y. (2006). A parallel and efficient algorithm for multicompartment neuronal modelling. *Neurocomputing, 69*(10–12), 1035–1038. doi:10.1016/j.neucom.2005.12.040.
201. Zienkiewicz, O. C. (1997). Trefftz type approximation and the generalized finite element method—history and development. *Computer Assisted Mechanics and Engineering Sciences, 4*, 305–316.

Index

© The Author(s) 2017

A.L. Madureira, *Numerical Methods and Analysis of Multiscale Problems*, SpringerBriefs in Mathematics, DOI 10.1007/978-3-319-50866-5